Smart-Tech Society

For Betsan

Smart-Tech Society
Convenience, Control, and Resistance

Mark Whitehead

Professor of Human Geography, Department of Geography and Earth Sciences, Aberystwyth University, UK

William G.A. Collier

Researcher, Department of Geography and Earth Sciences, Aberystwyth University, UK

Cheltenham, UK • Northampton, MA, USA

Cover image: Owen Beard on Unsplash.

Published by
Edward Elgar Publishing Limited
The Lypiatts
15 Lansdown Road
Cheltenham
Glos GL50 2JA
UK

Edward Elgar Publishing, Inc.
William Pratt House
9 Dewey Court
Northampton
Massachusetts 01060
USA

Paperback edition 2024

A catalogue record for this book
is available from the British Library

Library of Congress Control Number: 2022946661

This book is available electronically in the **Elgar**online
subject subject collection
http://dx.doi.org/10.4337/9781800884106

ISBN 978 1 80088 409 0 (cased)
ISBN 978 1 80088 410 6 (eBook)
ISBN 978 1 0353 4298 3 (paperback)

Printed and bound by CPI Group (UK) Ltd, Croydon, CR0 4YY

Contents

List of figures   vi
Acknowledgements   viii

1   The Smart-Tech Revolution   1

2   Analysing the smart-tech society   15

3   Prediction, personalisation, and the data self   43

4   Behaviour and freedom   75

5   The smart body—from cyborgs to the quantified self   99

6   Smart working and the corporation   127

7   Smart-tech states   148

8   Dumbing down—recalibrating our relations with smart
     technology   173

9   Conclusion   202

References   208
Index   217

Figures

2.1   Properties and processes affected at the interface of
humans and smart technology   19

2.2   Properties and processes affected at the interface
of humans and smart technology in relation to the
technological production of digital humans   28

2.3   HitchBOT   35

3.1   SenseMaker survey triad—reasons for using social media
platforms   53

3.2   Age profile of SenseMaker survey respondents   54

3.3   Educational achievement profile of SenseMaker survey
respondents   55

3.4   Narrative tone of SenseMaker story as determined by participants   55

3.5   Word cloud representation of terms used within participant
SenseMaker narratives   57

3.6   What benefits do you see in your social media platform
knowing you well?   65

3.7   How would you describe your use of this social media platform?   65

3.8   How did it make you feel when it anticipated your needs in
this way?   66

3.9   What do you attribute the predictive power of social media
platforms to?   69

3.10   What do you attribute the predictive power of social media
platforms to? The online cluster   70

3.11   What do you attribute the predictive power of social media
platforms to? The mixed cluster   70

3.12 The more a social media platform knows about me … 71

3.13 The more a social media platform is able to accurately predict my needs … 72

3.14 Behavioural response to accurate social media predictions 72

3.15 Change in usage over time following experience of social media predictions 73

5.1 Apple Watch sends its congratulations after I close my exercise ring 122

8.1 'How would you describe your use of this social media platform?' 186

8.2 'Using social media platforms facilitates …' 187

8.3 'Using social media platforms erodes my sense of …' 188

8.4 'What concerns do you have about your platform knowing you well?' 189

8.5 'What would your primary motivation be to use social media platforms less?' (Cluster 1) 189

8.6 'What would your primary motivation be to use social media platforms less?' (Cluster 2) 190

8.7 'How would you characterise your ability to stop using this social media platform?' (Cluster 1) 190

8.8 'How would you characterise your ability to stop using this social media platform?' (Cluster 2) 191

Acknowledgements

The research presented in this volume has been supported by two grants. The first was an Independent Social Research Foundation (ISRF) Mid-Career Fellowship: *Re-Thinking Freedom in a Neuroliberal Age* (2019). We would like to thank the ISRF for their generous financial assistance and for the broader support they offer the work of social scientists. The second was an Economic and Social Research Council [ES/P00069X/1] PhD studentship, *Corporate Governmentality*. Without the support of these organisations this volume would not have been possible.

We are also grateful to our colleagues in the Department of Geography and Earth Sciences and the Aberystwyth Behavioural Insights Interdisciplinary Research Centre at Aberystwyth University. Particular thanks go to Liz Gagen, Jesse Heley, Gareth Hoskins, Rhys A. Jones, Rhys D. Jones, Rachel Lilley, Pete Merriman, Sinead O'Connor, Mitch Rose, Marc Welsh, and Mike Woods. We would also like to acknowledge the support and inspiration offered by David Berreby, David Beer, Joram Feitsma, Matt Hannah, Rachel Howell, Matt Jarvis, Richard Marggraf Turley, Jessica Pykett, and Vranek (and his inspirational Vorsprung programme). Special thanks must go to David Snowden and Beth Smith at Cognitive Edge for their inspiration and support with our SenseMaker analysis. We presented aspects of this book to conference and seminar audiences at Bayreuth University, Goethe University Frankfurt, and Utrecht University. We are grateful for the comments and questions that those audiences raised.

We also acknowledge the support and encouragement of our family and friends, including Sarah Whitehead, Anwen and Betsan Whitehead, Janet Davies, Patrick Davies, Sue Whitehead, Carol Hawkins, and Steven and Helen Barnes. A special mention must also go to Elio Pace who has provided a constant source of musical inspiration during the long working hours that have gone into this book. Finally, we would like to thank all of the participants who took part in the survey and research interviews and shared with us their smart-tech experiences.

'Is it too late to dream of a modern society that has all kinds of wondrous technologies at its disposal and never ceases to invent new ones, that also remains committed to the question of what human life is for and how human beings should live, without allowing the answer to be technology?'

(Bruno Maçães, 2021: 24)

'Don't think too much about optimism and pessimism; try and catch the colour of life itself'.

(Henry James, 1884, *The Art of Fiction*)

1.   The Smart-Tech Revolution

The Neolithic period would certainly have been an interesting time to be alive. Looking back through the clouded veil of 12 000 years of human history, it is difficult to comprehend how profound the transformations of this time were. The gradual (it was very gradual) transition from gathering and hunting to more sedentary forms of life had deep and lasting consequences for human life. Homes, settlements, large families, domesticated animals, extended social networks, population growth, and cities were all flung into existence from this hinge of human history.[1] The Neolithic (or Agricultural) Revolution has been described as the first major disruption in human history (the second being the Industrial Revolution of the 18th and 19th centuries). The purpose of this book is to explore the lived experiences associated with what many believe is humanity's Third Disruption.

In his book *Fully Automated Luxury Communism*, Aaron Bastani explores the idea of a Third Disruption in human history. This disruption is marked by the rise of integrated circuits, digital technology, data circulation, and Artificial Intelligence (AI; Bastani, 2019: 37–49). It does, of course, go by many names: the Information Revolution, the Silicon Age, the Age of Automation, and, less optimistically, Surveillance Capitalism. Each of these monikers has its merits, but none quite convey the significance of the changes we are living through as well as the notion of a *Third Disruption*. The impacts of digital technology on human life include changing patterns of social interaction, knowledge acquisition, and learning; political participation; patterns of work; how we consume goods and services; and even how we make some of the most important life decisions (including who to date and where to live). The internet, smart homes (replete with digital home assistants and intelligent fridges), cyber communities and identities, virtual families and pets, machine learning and algorithms, social media networks, advances in biotechnology, collaborative robotics, workerless factories and lights-out manufacturing, and smart cities have all been flung into existence from this hinge of human history.

---

[1]   We acknowledge the misleading nature of the suggestion that it was only with the advent of sedentary agriculture that humanity saw significant social change. Evidence would now suggest that the world of hunter-gathering was one of 'bold social experiments' in human life and organisation (Graeber and Wengrow, 2021: 2–5).

It would be foolish to think that one book could make sense of all this social and economic change. This volume thus focuses on one dimension of this great disruption: smart technology. Smart technology is any form of digital instrument that is able to learn from previous actions in order to optimise future ones. This learning can take different forms. It could be Google's DeepMind neural network studying scientific databases to unlock the secrets of how proteins fold (and by extension, the key biological processes associated with cancers and ageing). Or Tesla digitally monitoring driver behaviour to inform the driverless car programmes of the future. This volume is primarily interested with how smart technology learns from and interacts with human users. It is unwise, however, to draw too strong a line of distinction between the learning that smart technology achieves through the monitoring of human behaviour and purer forms of big data speculation. These data sets and learning networks overlap in complex and often indecipherable ways. What ultimately unites these technological systems, and is a particular concern within this book, is that they inevitably extend the processes of knowledge accumulation, analysis, and learning beyond the biological limits of human perception, cognition, and understanding.

A concern with human interaction with smart tech draws our attention to particular domestic, everyday manifestations of such technology: searching Google, connecting with new friends on Facebook, changing our exercise patterns based on performance feedback from our Apple Watch, or setting up a Hive thermostat in our home. But this does not mean that we should lose sight of the broader infrastructures of digital surveillance that these technologies are embedded within (see Ball, 2020). In this context, we are not only interested in the physical devices that are commonly prefixed with the term 'smart' (including phones, watches, TVs, etc.), but also with the back-stage actions of algorithms, predictive analytics, machine learning, and AI, which all facilitate the learning associated with smart devices and platforms. In exploring the interfaces between humans and smart tech, it is also important to remember we are analysing a dynamic zone of socio-technological change. The capabilities of smart technologies are being constantly upgraded through ever greater exposure to the human condition. At the same time, what it is to be human is clearly being altered through exposure to the learning loops of machines and algorithms. The smart-tech–human interface involves adaptive change within smart systems and the rewiring of the human brain, behaviour, and social life.

There are many theories that indicate what human relations with smart technology means. At one end of this theoretical spectrum are more pessimistic theories, which suggest that we are already inescapably dependent on smart technology and the logics it is associated with. These theories often suggest that these dependencies will lead to a dark technological and social future associated with human disempowerment. At the other end of the spectrum are accounts of smart technology which suggest that it promises a bright future of efficiency, convenience, and a more luxuriant human existence. This volume is generally inspired and influenced by ideas and thinking that operate between these theoretical poles. Thus, while we are sympathetic to those who draw attention to the troubling potential of smart technology, we recognise that human relations with such technologies are far from set. This volume explores specific intersections of the human–smart-technology interface. In doing so it analyses the ongoing and unfinished socio-technological negotiations associated with the smart-tech society. Ultimately, we consider what these negotiations tell us about the human condition now, and what they may mean for our collective futures.

SMART TECH AND THE HUMAN CONDITION

Analysing the dialectical learning processes now connecting humans and smart technology raises questions that are fundamental to human life. It presents a mirror, a *Black Mirror*, to the human condition. In his 1911 book *An Introduction to Mathematics* the British-American philosopher Alfred North Whitehead observed, 'Civilization advances by extending the number of important operations which we can perform without thinking of them' (Whitehead, 1911: 46). Whitehead's observation is now supported by evolutionary psychology, which recognises that in the longue durée of human history the ability to conserve cognitive resources has been an advantageous trait. But, whereas in the evolutionary past not having to think too much was made possible by the actions of other, more instinctive, parts of the human operating system (particularly emotional and instinctive responses), now more of the cognitive load is being carried by technology. This raises the very real prospect of smart technology knowing us better then we know ourselves. A striking, and controversial, aspect of smart technology's potential in this regard can be seen in emerging forms of *relationship outcome predictions*. Relationship outcome prediction technologies utilise digital patterns in life-style activity to forecast the likely longevity of human relations. This means that Facebook and Alexa may know you are going to get divorced (at certain levels of statistical probability, at least) before you do (see Dickson, 2019).

Smart tech embodies a challenge to the human condition because it raises a question about what it is to be human. A defining characteristic of the human condition is self-awareness. Self-awareness and understanding are seen as *sine qua non* to the practices of self-determination and responsibility which have been the cornerstones of human civilisation for millennia. The idea that humans may not be best placed to know what they want, or need, or should do, represents nothing short of a foundational challenge to established political, economic, legal, moral, and even religious assumptions. Furthermore, if human decision-making is best outsourced to smart technology, what does this mean for tolerance towards alternative ways of living and more esoteric questions of human dignity?

The impact of smart technology on human decision-making is not limited to individual actions. Increasingly, smart tech is informing collective decision-making. The *smart-tech society* which this volume focuses upon should not be thought of as merely an aggregate of individual human actions. The social also pertains to how we orchestrate decisions at community, regional, national, and international levels. In this context, it is significant that algorithms are being used to make decisions on parole applications, students' likely examination grades, and even how to police protests (Amoore, 2020). In the UK, for example, the Home Office has recently been struggling to process the millions of visa applications coming from European Union (EU) nationals who wish to remain in the country following Brexit. It is claimed that in these situations smart technology offers enhanced processing power and supposedly greater impartiality. Concerns have been raised, however, that government algorithms just hard-wire biases (particularly in relation to age and nationality discrimination) into collective decision-making systems (Warrell, 2019).

Whether at an individual or collective level, smart technology is raising unsettling questions about what it is to be human and about established social norms and practices. As we develop and deepen connections with smart technology, more of what makes us human—including our memories, our sense of responsibility, how we choose, and what we learn—will be the product of novel amalgams between *Homo sapiens* biology and digital technology. Such developments are already transforming the social realm—including how we meet, deliberate, and make collective decisions. It is also abundantly clear that we are at the thin end of this technological wedge.

LUXURY TRAPS AND REVOLUTIONS: THE LESSONS OF THE THIRD DISRUPTION

We now return to our Neolithic ancestors going through their collective revolution. We do so because the parallels between the Agricultural Revolution and our neoliberal digital transformations are more nuanced than the fact that they are both 'big and historically rare things!' Furthermore, these parallels can help us begin to assess the beneficial trajectories of the Third Disruption, and the traps that may be found within it.

In the popular consciousness, it is rare to think of the changes associated with the Neolithic Agricultural Revolution as anything but positive developments. It is established wisdom that the First Disruption enabled humans to move from a Hobbesian *state of nature* and to live lives that were less solitary, less poor, and generally less nasty, brutish, and short.[2] In his analysis of the Agricultural Revolution, Yuval Noah Harari suggests a different interpretation. According to Harari, 'Village life certainly brought the first farmers some immediate benefits, such as better protection against wild animals, rain and cold. Yet for the average person, the disadvantages probably outweighed the advantages' (2014: 93). It is perhaps ironic that despite producing more food for human consumption in the aggregate, for many, the emergence of sedentary agriculture led to more disease (a product of settled communities), greater malnutrition (due to the devasting impacts of crop failures), and a far less interesting diet. For Harari then, the First Disruption 'offered little for people as individuals' (2014: 93). Through the intensified production of food per spatial unit, the revolution was, however, able to 'keep more people alive under worse conditions' (2014: 94). While not benefiting individual farmers, this was a huge boon for the human species and its DNA code.

We do not use the example of the Neolithic Revolution simply to make the obvious point that the Third Disruption may not be the panacea some see it to be. Our aim is rather to suggest that, in contrast to the Neolithic era, the Third Disruption generally brings clear benefits to the individual, but disadvantages to the human species as a whole. The advantages are clear and evident in our everyday lives. Smart technology brings convenience, speed, personalisation, new opportunities for social connection and matching, novel platforms for bespoke forms of self-promotion, and useful forms of feedback and social comparison. There are, of course, disadvantages to smart technology, which are also obvious in quotidian contexts. These include the addictive qualities of social media, identity theft and fraud, and the loss of privacy. But many of the broader disadvantages, relating to behavioural manipulation, human autonomy, a loss of political control and accountability, and digital surveillance, are less apparent (at least in terms of their full implications). The immediate benefits of the Third Disruption to people may, in part at least, explain its rapid onset. But even if the advantages of the Third Disruption are more immediately obvious than those of the Neolithic Revolution, and its disadvantages are far less apparent than its historical predecessor, there is still much we can learn from the First Disruption.

The First Disruption may provide us with insight into why humans subject themselves to long-term disadvantages, and the challenges that may exist to reversing change once it has been set in motion. For Harari (2014), the Neolithic Revolution was ultimately a bargain with wheat. The cultivation of wheat enabled humans to sustain larger populations. These larger populations

---

[2]   Although again, for an alternative perspective, see Graeber and Wengrow (2021).

were in turn needed to sustain the back-breaking commitments that were associated with the production of more wheat. Harari observes that '[p]ara-doxically, a series of "improvements", each of which was meant to make life easier, added up to a millstone around the necks of farmers' (2014: 97). In this volume we are interested in whether a similar paradox may be associated with the Third Disruption. Could the 'improvements' associated with smart tech be contributing to a collective millstone around the necks of 21st-century humanity? If history is repeating itself, the Neolithic Revolution may provide insight as to why. In this context Harari asks the important question of why the people of the Neolithic period actively made their lives worse. He argues that this collective miscalculation was a product of the fact that '[p]eople were unable to fathom the full consequences of their decisions' (2014: 97). It certainly seems clear that we are not able to comprehend the full implications of the Smart-Tech Revolution. Could we be on the cusp of a collective miscal-culation? And if we are, is there anything we can do about it?

The purpose of this volume is not to merely chart the potentially disastrous consequences of the Third Disruption. We remain genuinely open to the pros-pect that the smart-tech society could be a harbinger to a great age of human luxury, plenty, and opportunity. We also recognise and reflect upon the numer-ous ways in which smart tech already makes our lives better. We do assert, however, that just like our ancient ancestors, as participants of the smart-tech society we are unable to understand the full consequences of our decision to abdicate certain aspects of our decision-making to digital technology.

The Neolithic period also indicates the likely irreversibility of the com-mitment to smart technology. Despite the travails associated with sedentary agriculture, it appears that Neolithic people could not simply choose to return to their hunter-gathering ways. By committing themselves to a settled life, wheat growing, and larger populations, early agriculturalists essentially burned their boats: or to put things more bluntly, as Harari (2014) does, to return to hunter-gathering, humans would have had to choose to go through a period of extreme food shortages, famine, and population decline. While the consequences of reversing the Third Disruption may not be quite as dramatic, as we allow smart technology to influence more of our lives, unplugging tech becomes ever more difficult.

Interestingly, Harari draws a direct line of comparison between the Neolithic period's dependence on wheat and modern society's reliance on technology. He claims that in the contemporary pursuit of an easier life our reliance on technology could see us fall into a luxury trap: 'One of history's few iron laws is that luxuries tend to become necessities and to spawn new obligations' (2014: 98). In many ways the notion of a luxury trap is a helpful metaphor when attempting to interpret the nature of the smart-tech society. It is clear, for example, that the appeal of smart technology derives from the labour-saving luxuries (both manual and cognitive) it promises. It is also clear that many forms of smart technology have already transitioned from being luxuries into necessities. But this metaphor also has limitations. The notion of luxury suggests that the uptake of smart tech is largely about opting for convenience. Yet, as we will see, smart technology is not always something we choose to adopt (even in its early luxury phase). Neither it is something that necessarily makes our life easier. In this context, the idea of a luxury trap should be thought of in more than metaphorical terms. As a metaphor, the luxury trap suggests something we become ensnared in accidentally. But as this volume will demonstrate, there are more deliberative attempts to spring the traps of smart technology currently at work. Just like the First Disruption, nobody knows precisely where the Third Disruption is taking us. Unlike the Neolithic period, however, our transition towards a smart-tech society is far from accidental.

THE SIGNIFICANCE OF 2008

To gain a better sense of the historical significance of the smart-tech society we are building, it is helpful to return to 2008. For many, 2008 will forever be remembered as the year of the Credit Crunch and the political and economic turmoil it generated. At some point in 2008 (perhaps 2009), however, a much less discussed but perhaps even more significant event occurred. This was the moment that, for the first time in human history, there were more devices connected to the internet than there were people on the planet. There are several things that are of note about this moment in time. First, is the rate of growth associated with internet-connected devices. In 2003, there were an estimated 500 million such devices; by 2008 this number had grown to in excess of 6 billion (that is a 12-fold increase in the space of 5 years!; Evans, 2011). But it is only after 2008 that the numbers get really interesting. According to Cisco, by the early 2020s there could be as many as 50 billion devices communicating with the internet: that is an approximate average of seven devices per person on the planet (Evans, 2011)!

Taken together, this growth of internet-connected devices is often referred to as the *Internet of Things*. The Internet of Things (or IoT) is the expanding array of devices that are connected to the internet. Many of these devices—such as smartphones or laptops—will have an associated user. But many others operate independently of direct human input. The IoT now includes network-connected hairbrushes, mirrors, rectal thermometers, fridges, and even concrete sensors (Burgess, 2018). Many of these devices are relatively dumb when operating in isolation. When they share information with other devices, however, they take on smarter forms. It is often unclear what the utility of many of the devices that constitute the IoT actually is, or indeed why a light bulb needs to be able to communicate with the internet. But what the rise of the IoT appears to reflect is an historical tipping point. A point at which our collective digital existence is defined less by humans searching the internet, and more by the internet searching us!

The expansion of the IoT is being driven by a series of processes. First are attempts to upgrade mundane objects, such as fridges or doorbells, to boost the sales of goods and services. Second, for Big Tech the growth of the IoT directly benefits their business model by facilitating the gathering of ever more user data for analysis and sale. There is a third reason for the growth of the IoT though; this is the one that we find most fascinating. It has been claimed that the growth of the IoT is a kind of open experiment without a clear end goal (see Zuboff, 2019). Given how cheap the sensors associated with smart technology tend to be, digital monitoring is often being expanded for no other reason than it is easy to do. This form of experimentalism is typical of the permissionless innovation that has defined the digital era (Barbrook and Cameron, 1996). According to Zuboff (2019), the growth of the IoT represents a form of applied utopianism. Utopianism generally offers the promise of a better future whose basic outline can be described. Conventional utopian visions have the advantage of enabling people to question whether the societies that are promised are the ones they want to live in. Applied utopianism, on the other hand, promises a brighter future but offers no clear outline of what that future will look like. It is a form of *applied* utopianism because its realisation can only be achieved through its iterative construction in the here and now—there is no blueprint to work from. In reality, all utopias are applied, at least to the extent that their final form is shaped by what the practical working through of ideological visions involves. But, the smart-tech society offers no clear vision of where we are collectively going, only a commitment to the fact that we need more smart tech in order to build its eponymous world. The smart-tech society is thus not only characterised by a general inability to fathom the kind of world we are collectively submitting to. Its primary architects do not even know where we are heading. A primary concern of this volume is how we individually and collectively respond to this indeterminacy.

TECHNO-UTOPIANISM AND DYSTOPIA: FROM CYBERDYNE TO GAIA 2.0.

OK, so we don't actually know where we are headed. This, of course, has not prevented people from speculating about the likely form our smart-tech society will eventually take. Some visions of the smart-tech society predate our encounters with smart tech as we currently understand it. Kubrick's chilling depiction of HAL, Asimov's laws of robotics, and the techno-violence of Cameron's Cyberdyne Systems all anticipate human encounters with technological intelligence. Indeed, much—if not all—of the sci-fi genre anticipates the issues the smart-tech society presents us with in one way or another. Although we may not know where the smart-tech society will end up, in some ways our technological present is already the science fiction future of the past. It is thus interesting to consider which fictional dystopian tropes appear to anticipate our current situation most accurately. A common science fiction theme is the notion of a loss of human control over sentient creations. This can, of course, be seen in what many consider to be the first work of scientific fiction, Mary Shelley's 1818 *Frankenstein* (although more on this in a moment). In later works of fiction, it is the cyborg (cybernetic organism) who tends to occupy the centre ground of dystopian tales. In the figure of James Cameron's Terminator, the cyborg is an organic expression of an all-conquering AI system using its superior intelligence to push humanity to the edge of extinction. In Philip K. Dick's *Do Androids Dream of Electric Sheep?*, the human–technology hierarchy is reversed, as it is cyborgs (or replicants) who are hunted down by human authorities as they desperately seek more life. The cyborg metaphor becomes most interesting in Dick's novel. Here (and in its postmodern retelling in Ridley Scott's 1982 cyberpunk film *Blade Runner*) the boundary line between androids and humans becomes blurred and a sense of ambiguity is cast on the distinctions between the human and the technological (Bukatman, 1997).

Ultimately, however, the corporeal figure of the manufactured cyborg appears to miss the mark of what the smart-tech society looks like. This is not to say that the metaphor cannot be an interesting and useful one (see Chapter 2 of this volume for more on this point). Our smart-tech society appears, at present at least, to be taking a different form. Wearable smart technology does mean that humans are increasingly taking on a cyborg aesthetic, but our emerging relations with digital technology are more diverse and complex than can be captured in the corporeal figure of the cyborg. There are three further fictional portrayals of the technological future, which perhaps takes us closer to the nature of the smart-tech society we are confronting. The first also originated in the mind of Philip K. Dick—the short story *The Minority Report*. In this 1956 novella, Dick describes the existence of mutants, 'pre-cogs' who are able to predict crime before it occurs. When plugged in to a pre-crime policing machine, they support the exercise of pre-emptive policing and prosecution. Dick's pre-crime world of *The Minority Report* raises many concerns that are of direct pertinence to the smart-tech society. The anticipatory qualities of Dick's pre-cogs echo the predictive analytics that are being deployed within smart-tech systems. It is also clear that smart tech is already informing criminal justice systems—through algorithmic understandings of crime (Završnik, 2021)—and the pre-emptive policing of protests (see here Amoore, 2020). But it is in relation to questions of free will and autonomy that *The Minority Report* speaks most pressingly to the problematics of the smart-tech society.

As we discuss later in this volume, the smart-tech society is characterised by an emerging relationship between data science and behavioural science. This fusion means that not only can smart tech predict our needs and desires, but through behavioural prompts, it may hasten related actions into being (see Chapter 4 of this volume). The closing of the gap between desire and action has strong parallels with the pre-policing of crimes as described in *The Minority Report*. In closing the gap between desire and action, the predictive systems of the smart-tech world may make our lives easier, but they also pose a challenge to our collective decision-making rights. They also challenge established understandings of autonomy and responsibility that are central to many existing social institutions.

The second depiction of the technological future that is of particular significance to this volume is the 1999 sci-fi action film *The Matrix*. In this popular film, directors Lana and Lilly Wachowski portray a future in which humans are unwittingly plugged into a virtual-reality world, while their bodies are used to power a machine world of superior intelligence. Despite its extreme dystopian depiction of human–technology relations, the world of *The Matrix* has key parallels to the smart-tech society of today. The idea of intelligent technologies wilfully distracting humans reflects the addictive attention-based economies associated with the Third Disruption. From the Facebook 'like' button, to personalised content and arresting news feeds, smart tech is increasingly geared towards capturing more and more human attention. Interestingly, the virtual reality of *The Matrix* has also been tweaked to be imperfect (the perfect worlds of the matrix's early manifestations were rejected by humans' collective subconscious). Here we find clear parallels with the iterative, and potentially infinite, trials that smart technology carries out on humans to determine their true desires and preferences (desires and preferences that humans themselves may not be completely aware of). The final aspect of *The Matrix* that is of significance to this volume is the mutual dependence it depicts between people and machines. This is not a story of cyborgs seeking to destroy humans, or humans eviscerating replicant rivals. It is an account of mutuality: of smart technology that needs humans, and humans who are in turn dependent on technology for their virtual (and underground) existence.

We argue that it is fictional accounts of the ambiguous dependencies that exist between humans and technology that are most relevant for the present time. It is precisely these forms of ambiguous dependency that can be found in Shelley's account of Dr Frankenstein and his monster. The monster is both a threat to and protector of humans; it seeks isolation from humans while desiring their company and interaction. We argue that it is these forms of socio-technological ambiguity and equivocality that characterise the Third Disruption. We hope this volume can offer an academic account of some of the negotiated uncertainties of the smart-tech society that reflect the interpretative qualities of these fictional stories. This commitment to ambiguity is perhaps best thought of in methodological terms. We thus see this commitment not as a white-flag-waving exercise: using it to lazily claim that reality is simply too complex to definitively analyse. Instead, we see it as a methodological commitment to paying attention to the operative function and essence of smart tech, rather than axiomatically explaining it through existing interpretative frames.

*

There is, of course, no shortage of utopian depictions of the digital, smart-tech future: Google's promises to benevolently order global knowledge, Facebook's desire to connect the global population, and Bastani's (2019) aforementioned accelerationist vision of fully automated luxury communism. What we are primarily interested in in this volume is a deeper, often hidden, form of utopianism that characterises the smart-tech society. To understand this utopianism we have to go back to a rather unlikely starting point: the work of the American-Russian writer and philosopher Ayn Rand. Rand is best known for her popular novels *The Fountainhead* and *Atlas Shrugged*, and her related philosophy of objectivism. Throughout her work, Rand emphasised the importance of reason and ethical egoism. At the root of ethical egoism was a belief in the value of pursuing self-serving rational projects and rejecting the inhibiting commitments of altruism. Rand's books are characterised by heroic individualists who pursue their visions against oppressive social and govern-ment systems. Although Rand's work does not anticipate a smart-tech society, it did inform the thinking and practices of its key figure in the Big-Tech sector. Rand's novels and ideas were prominent influences in the entrepreneurial community of Silicon Valley in the 1970s and '80s (Curtis, 2011). Many smart-tech entrepreneurs saw themselves as Randian heroes, whose pioneering developments had the potential to disrupt established social hierarchies and oppressive social systems. It was these utopian principles that informed the ideas of permissionless innovation that continue to inform the Smart-Tech Revolution (Ohanian, 2013). It is these ideas that continue to inform the fearless pursuit of what smart technology can offer despite its evident dangers. These ideas influenced John Perry Barlow when, in Davos in 1996, he pub-lished the *Declaration of the Independence of Cyber Space*. In this declaration Barlow (1996) outlines his provocative utopian vision of a future free of the cumbersome regulations of the industrial world,

> Governments of the Industrial World, you weary giants of flesh and steel, I come from Cyberspace, the new home of Mind. On behalf of the future, I ask you of the past to leave us alone. You are not welcome among us. You have no sovereignty where we gather … I declare the global social space we are building to be naturally

independent of the tyrannies you seek to impose on us. You have no moral right to
rule us nor do you possess any methods of enforcement we have true reason to fear.

The global (cyber) social space described by Barlow is not a blueprint for
our smart-tech society. As we have already stated, no such blueprint has ever
existed—it has always been an applied utopianism. But it is this spatial vision
that has enabled the smart-technology world of today to evolve in the way that
it has.

Barbrook and Cameron (1996) suggest that the fusion of Randian ethics and
technological utopianism ultimately produced a distinct ideology. They named
this the Californian Ideology to reflect its West-coast origins, but it could just
as easily be called the San Francisco Bay Ideology. According to Barbrook
and Cameron the Californian Ideology is a kind of heterogeneous orthodoxy that is riven with peculiar contradictions. It fuses the left-wing hippy
anarchism of 1960s San Francisco with the economic liberalism of Silicon
Valley. Added into this incongruity is a heavy dose of techno-determinism.
Drawing inspiration from Marshall McLuhan's vision of the Global Village,
this determinism depicts an inevitable drift to a future built on West-coast
technologies. The California Ideology combines an anti-corporate ethos of
the gift economy (which can still be seen in the free service model of many
high-tech giants), with an anti-state rhetoric that echoes Barlow's digital declaration of independence (which can still be seen in the anti-regulatory stance
of the smart-tech industry). As a utopia of both the political left and right, it is
perhaps deserving of the title of the true Third Way.

What is interesting about the Californian Ideology is that if it is a utopian
vision, it is as much a utopia of the present as it is of the future. It embodies
the technological basis for the neoliberal dreams of Reagan and Thatcher.
Neoliberalism suggests that it is only through the market that society can have
personal freedom and social stability. But it is the Californian Ideology which
ushers in the digital space that can enable the flow and monitoring of the
copious forms of market information that are needed to theoretically deliver
the neoliberal ideal of self-regulating market stability (Curtis, 2011). The
digital realm is thus seen as the natural environ of the market: information-rich
and monopoly-free. As a utopia that is home to both hippies and yuppies,
that combines technological determinism and individual liberation, and even
manages to exhibit a romantic Jeffersonian feel, the Californian Ideology was
always too good to be true (Barbrook and Cameron, 1996). But, as with all
ideologies, it produced an operational belief system. This system of belief
suggests that a Randian amoral individualism could be combined with broader
forms of social stability that were enabled through ease of interaction and the
free flow of market knowledge. The failure of this belief system is evident in
the crisis of neoliberalism today. But, as we will see, the emergence of smart

technology offers the basis for the resurrection of a super-charged Californian Ideology, within which the free flow of market information is replaced by total market knowledge, algorithmically perfected decision-making, and behaviourally adjusted market actors.

The final smart-tech utopia we would like to reflect upon is, perhaps, the most surprising. It is certainly the most far-reaching. It was developed by the eminent NASA scientist and world-renowned author James Lovelock. In the 1970s Lovelock advanced his theory of Gaia: the idea that the Earth is an integrated environment system best thought of as a self-regulating, super-organic entity. While often dismissed at the time by the scientific community, these ideas provided the basis for contemporary Earth System Science. In his 2019 book *The Novocene: The Coming Age of Hyperintelligence*, Lovelock considers the implications of smart technology for the planetary system. Lovelock's analysis radically ups the stakes of the smart-technology debate, interpreting the coming smart-tech society in evolutionary and geological terms. In geological terms the Novocene is depicted by Lovelock as a new epoch that is defined by smart technology (a kind of age of cyborgs). Lovelock does not see this as merely a significant development in human history, but something of planetary significance. For Lovelock, the Novocene is the Earth's Third Disruption, following the appearance of photosynthetic bacteria 3.4 billion years ago, and the Industrial Revolution of the late 18th century (2019: 87). According to Lovelock, the Novocene is of particular importance because it will usher in the end of the Anthropocene. The Anthropocene is a contested geological designation that depicts the deleterious impacts humans have had on the global environment. In Lovelock's (2019) estimation, then, it is smart technology that will save the planet:

> Live cyborgs will emerge from the womb of the Anthropocene ... it is crucial that we should understand that whatever harm we have done to the Earth, we have, just in time, redeemed ourselves by acting simultaneously as parents and midwives to the cyborgs. They alone can guide Gaia through the astronomical crises now imminent. (86)

The cyborgs of Lovelock's Novocene (he envisages them as wise spheres) will apparently be benevolent. According to Lovelock their priority will be stabilising the conditions for life on Earth, which will be optimally achieved by working with humans. While the Novocene is thus the twilight of humanity, it is a peaceful one, more retirement than evisceration by Cyberdyne Systems.

The key epistemological shift associated with the Novocene is a move from human- to planet-centred action. Lovelock thus observes, 'By giving computer systems of the Anthropocene the chance to evolve themselves by natural, or assisted selection, we take away the barriers that impeded Gaia's move to its next state, the Novocene, where self-regulation is no longer aimed at support-ing our form of biosphere alone' (2019: 114). The evolutionary shift depicted here is from one of natural to intentional selection. This not only suggests a change in the nature of evolution, but also in the function of smart tech. No longer is smart tech about human-assisted thinking; it is about transcending humanity. While Lovelock's vision of Gaia 2.0 may seem far-fetched, it is worth emphasising that this is not emerging from the mind of a science fiction writer. It is the product of a thought experiment pursued to its logical end by an eminent scientist. Furthermore, Lovelock's vision of hyper-intelligence put to the service of smart geo-engineering is evident in many of the existing ecological claims we find in the contemporary smart-tech industry.

We do not share Lovelock's surprising optimism about the nature of hyper-intelligent technology and its implications for the human future. His geological level of analysis also takes us away from the more mundane domes-tic focus of this volume. His speculations do, however, serve to emphasise just how significant smart tech could be. His work also raises important questions about the relationship between humans and smart tech. In his depictions of humans as the parents of smart technologies that will ultimately serve as benevolent carers for humanity's home, Lovelock opens up some interesting ways of conceiving the human–smart-tech dialectic, and the learning systems it embodies. We will return to these questions throughout this volume.

STUDYING THE SMART-TECH SOCIETY: CONVENIENCE, CONTROL, AND CONNECTION

The aim of this chapter has been to set the broader historical and speculative scene for our analysis of the smart-technology society. We have said relatively little of our specific approach to the study of smart technology. This approach is detailed in the following chapter. In the next chapter we position our own analysis within a broader set of approaches to the analysis of human interac-tions with technology. As we have hinted at in this chapter, our approach is defined by a commitment to studying the ambiguities, uncertainties, and messy compromises that characterise human–smart-tech interactions. Our approach is a composite of existing approaches to related questions. Rather than ascrib-ing to an established intellectual narrative, or utopian/dystopian thinking, our analysis is primarily informed by an attentiveness to various empirical interfaces of human and smart technologies. We thus seek to make sense of the smart-tech world not by speculating about its likely future form, but by paying attention to what it is like to live in it now.

2. Analysing the smart-tech society

INTRODUCTION—THE END OF THEORY

In 2000, the editor of *Wired*, Chris Anderson, wrote an article for the magazine titled 'The End of Theory: The Data Deluge Makes the Scientific Method Obsolete'. Anderson argued that as data gathering and storage enter the *Petabyte Age* they present new epistemological challenges and opportunities to humankind: 'The Petabyte Age is different because more is different. Kilobytes were stored on floppy disks. Megabytes were stored on hard disks. Terabytes were stored in disk arrays. Petabytes are stored in the cloud' (Anderson, 2000).

The scale of data gathering in the Petabyte Age (a petabyte is one thousand million million bytes, in case you are wondering) is a product of enhanced data storage technology but is also driven by the digital data gathering and processing requirements of smart technology. For Anderson, this scale of data does not just present technological challenges for humankind, it also challenges our epistemological practices as we strive to make sense of the world. According to Anderson (2000), the first epistemological implication of the Petabyte Age is the pre-eminence it gives to mathematical techniques of analysis:

> At the petabyte scale, information is not a matter of simple three- and four-dimensional taxonomy and order but of dimensionally agnostic statistics. It calls for an entirely different approach, one that requires us to lose the tether of data as something that can be visualized in its totality. It forces us to view data mathematically first and establish a context for it later.

Anderson suggests that within the digital age human data-gathering capacities inevitably go beyond the human capacity to adequately comprehend the nature of those data. Mathematics, and not human perception, is thus given epistemological priority in the smart-tech society.

In this volume we are conscious of the need for mathematical processes to make sense of the data deluge. In this chapter (and the wider book), however, we reassert the value of the qualitative study of human experience within smart-tech systems. We acknowledge that much of the nature of, and the epistemological perspectives produced by, smart-tech systems are beyond human perceptive capacity. We nevertheless argue that it is impossible to understand the continuing development of smart-tech systems, and wider digital worlds, without a grasp of the human experience of them. Although the Petabyte Age may produce data at scales and levels of complexity that are incomprehensible to human senses, this does not mean that all that can be known about this new world is in the mathematical register. In this sense it could be argued that we are interested in the human experience of the mathematical ordering of digital life.

The second implication of the Petabyte Age identified by Anderson (2000) relates to questions of theory. Throughout large swathes of human history theory building enabled humans to look at fragments of reality and to speculate what these fragments tell us about the broader systems of which they are a part. For Anderson the Petabyte Age challenges this conventional model of theory building. Reflecting on the example of Google, Anderson (2000) states,

> Google's founding philosophy is that we don't know why this page is better than that one: If the statistics of incoming links say it is, that's good enough. No semantic or causal analysis is required. That's why Google can translate languages without actually 'knowing' them (given equal corpus data, Google can translate Klingon into Farsi as easily as it can translate French into German).

Ignoring the evident limitations in *Google Translate*'s ability to accurately decipher languages (this is, at least, our experience with Welsh), Anderson's comment draws attention to an important epistemological dimension of the smart-tech society. Due to the data deluge, smart platforms don't have to discern a theory of causality, which can then be tested in other contexts. Why theorise when Big-Tech can just know what is happening? The smart-tech world's apparent desire to transcend theory has significant implications across the social sciences: 'Out with every theory of human behavior, from linguistics to sociology. Forget taxonomy, ontology, and psychology. Who knows why people do what they do? The point is they do it, and we can track and measure it with unprecedented fidelity. With enough data, the numbers speak for themselves' (Anderson, 2000). Zuboff (2019) describes this approach to knowledge acquisition (and the power that derives from it) as *instrumentarianism*. This admittedly cumbersome term offers a critical interpretative framework to position smart-tech society's epistemological systems. According to Zuboff (2019), *instrumentarianism* embodies

> [a]n unusual 'way of knowing' that combines the 'formal indifference' of the neo-liberal worldview with the observational perspective or radical behavioralism … instrumentarian power reduces human experience to measurable observable behavior while remaining steadfastly indifferent to the meaning of that experience. I call this new way of knowing *radical indifference*. It is a form of *observation without witness*. (377, emphasis in original)

Replacing theory with a totalising vision of knowledge has significant implications for the place of humanity within the smart-tech society's epistemological system. For Zuboff, it implies that mathematical knowing can replace the meaning that humans place on observed phenomena. For Anderson, it also means that we need concern ourselves no more with the inner logics of why people do what they do. As a form of observation without witness this system of knowledge not only seeks to move beyond concerns within human motivation and meaning; it also displaces humanity as the locus of analytical scrutiny of the world.

*

Later in this volume we will consider the actual effects this post-theoretical prioritisation of correlation over causality has for the human condition. In this context, we draw particular attention to the connections between smart technology and behaviouralist approaches to human conditioning (see Chapter 4). But for now, we use Anderson's call for the end of theory as a counterpoint for the overtly theoretical purpose of this book. In this chapter we argue that despite smart-tech's tendencies towards dehumanised forms of observation and analysis, it is now more than ever that we need theories of the smart-tech world. Despite its knowledge claims, it is clear to us that in its abandonment of the human perspective, smart-tech systems themselves are uniquely unqualified to tell us about the human experience of such technological systems. Smart-tech systems claim to know all that needs to be known about human experience. We claim, however, that by focusing on only that which is observably measurable that such systems, and the Big-Tech interests they serve, are ignoring a significant pool of human experience. This is likely to be an experiential perspective that is important in shaping the future form of the smart-tech world. If we know more than we can say, then surely there is much more to know than can be discerned in our observed behaviours (Pasquale, 2020: 24).

Furthermore, we argue that proclaiming an agnostic position of observation without witnessing smart-tech systems is unhelpfully obscuring the vested interests that are informing digital knowledge-production systems, and limiting other ways of knowing the world. In this context, attempts to inhibit a diversity of analytical perspectives on the smart-tech society has political as well as epistemological implications. As we outlined in Chapter 1, this book has not been written as a volume that is inherently for or against a smart-tech society. What it is against, however, are attempts to use the knowledge gathering and analysis capacities of smart-tech systems to foreclose different perspectives on that very society. The smart-tech society is an historically unprecedented yet still-emerging socio-technical system. As a diverse collection of people, economic transactions, political regulations, technological devices, codes, and algorithms, it is precisely the type of underdetermined system that we need to build theories about. Without such theories we have no real hope of making sense of this complex and obscured utopia, or of hoping to engage in a democratically informed discussion of the future forms we may wish it to take.

To these ends, this chapter offers two things. First, it provides an overview of the critical perspectives that inform the empirical chapters that follow. Second, and unlike most theoretical chapters, it can also be read as a thematic chapter: exploring the connections among smart-tech, humans, and knowledge. The chapter has three main sections. First, it outlines the primary focus on this volume—namely, the relationship between smart tech and humans. Second, it explores the main theoretical frameworks that are available to help us make sense of this relationship. Finally, we outline the critical phenomenological framework of analysis within which we position our own analysis.

PERSPECTIVES ON THE HUMAN–SMART-TECH INTERFACE

Introducing the Interface

As we briefly outlined in the previous chapter, the primary focus of this volume is the various intersection points that exist between humans and smart technology. It is important to recognise, however, that to speak of the interaction between the two separate spheres of 'humanity' and 'smart technology' is misleading. Although computer programmes can now 'automatically' generate their own programmes, smart technology is a human creation and for the foreseeable future will be dependent on human inputs in various forms. At the same time, many of us are now reliant on smart technology for various aspects of our personal, social, and working lives—smart tech is already shaping and augmenting our humanness. The focus of this volume is thus more precisely understood as the various aspects of humanness, which smart tech is likely to influence (Figure 2.1). We are, however, also concerned with the ways in which individual and collective human actions can disrupt and change the evolving nature of smart technologies.

This stated concern does pose another question: why focus specifically on smart technology as opposed to more commonly explored processes of automation, algorithms, AI, or robotics? The primary advantage of focusing on smart technology is that it spans issues related to all of these things. Analysing smart technology does, to some extent, always involve aspects of automation: smart technology is smart to the extent that it seeks to make human decision-making easier through the production of automatically produced options and actions, or by replacing human labour. But the focus on smart tech is different from a more general focus on automation because smart-tech automation is derived

Figure 2.1　　Properties and processes affected at the interface of humans and smart technology

from machines learning from human action. For example, driving an automatic car may make life easier, but it does not offer insights into the emerging learning cycles that are being forged between humans and digital technology. This volume will, from time to time, reflect on questions of robotics. Robots, as programmable devices, often do incorporate smart learning capacities. Robots are also of interest because of the ways in which they are often designed to reflect human norms and values. But a focus on robots as task-oriented devices precludes the complex networks of monitoring devices and wearable technologies that constitute the smart-tech world. Finally, while algorithms, AI, and associated machine learning systems are characteristics of all smart technologies to some extent, the algorithm and AI are not how the smart-tech–human interface is recognised and experienced by people. While this volume is interested in the connections between people and smart technology in general terms, it is primarily concerned with the human experience of these interfaces. As such, the notion of smart technology provides a socially relatable context within which to explore the often-hidden impacts of algorithmic processes.

Figure 2.1 indicates some of the issues that will be explored in this volume. This list of themes demonstrates the great diversity that now characterises people's relations with smart technology. It also demonstrates precisely what is at stake within the analysis that we present here. Smart technology has implications for many of the biological, psychological, and social processes that are intrinsic to the human condition. Smart technology through various forms of e-education is now shaping how we learn about ourselves and others. Through algorithmically ordered searches and news feeds, smart tech is also shaping the ways in which we perceive the wider world and the judgements that we make about it. Through its demands on our attention, smart technology is also reshaping the ways in which we relate to those in our family and wider social circles. Smart technology also has implications for parenting techniques and our working practices, not to mention more existential questions of truth and trust. As indicated in Figure 2.1, our focus in this volume is both on the new capabilities that smart technology can offer the human condition and the novel challenges it generates. In terms of parenting, for example, this means considering how smart technology can assist in the diverse practices associated with child rearing (perhaps through the building of support networks, the provision of guidance on health matters, choosing a school, and educational tools), while at the same time generating new parenting challenges (particularly how to deal with addictive digital behaviours among children). At the centre of our analytical project, however, remains one question: How is smart technology changing the human condition? If being human is associated (no matter how tenuously) with the holding of certain memories, wielding the autonomous capacity to act, and being able to make independent judgements, what does the transactive movement of some of these capacities into the smart technological realm mean for humanity? For some, this will inevitably result in a diminution in the human condition, as we become what Carr (2016) describes as passengers in the smart-tech society. For others, the smart-tech society is associated with the augmentation of human capacities, an enhancement of our capacities to be who we would ideally like to be, and the provision of more time to do what we want to do.

As we established in Chapter 1, there are existing optimistic and pessimistic perspectives on smart technology. Our focus on the human experience of smart technology means that we endeavour to avoid an axiomatic reliance on existing utopian or dystopian interpretative frameworks. Our analysis of human–smart-tech interfaces thus aims to produce diagnostically specific accounts of particular interactions between people and digital technology. A helpful example of the style of analysis adopted within this volume is provided by Munger's (2021) reflections on the impacts of smart technology on human communication. Munger is interested in the contrast between social communication and the forms of communication that are favoured by machines. Social communication (in written and oral form) is about more than conveying information. We convey so much more when we speak or write than simple information sharing. For Munger (2021), 'these exchanges are social—they make sense only as a way of treating the other as a person rather than as a source of information'. In contrast, Munger recognises that machinic communication is about the efficient transfer of a clear signal. This contrast between human and algorithmic communication becomes significant in rela-

tion to the use of smart technologies, such as Gmail's predictive text and suggested phrases. While seemingly innocuous, Munger (2021) argues that such
communicative technologies may have long-term implications for the nature
of human communication and how we interpret the messages and signals we
receive. Through their ability to suggest clearer forms of human expression
(ones that their training data suggest have been most effective), these predictive technologies could make written communication more functionally
efficient. By removing social context (and presumably unnecessary preambles
and adjectives), machine-augmented technology can help remove ambiguity
from communication and enhance its clarity. But Munger also considers the
human cost of incorporating smart technology into our social communication
system. Munger (2021) claims that if humans start to communicate more like
algorithms, then social communication will inevitably be devalued. There
will be less opportunity to communicate feelings and interpersonal values in
messages. These seemingly human irrelevances play a vital role in our social
worlds, providing encouragement, guidance, and purpose within our interactions. Munger (2021) thus expresses concern that over time, such digital
patterns of communication could also start to change offline conversations and
their wider social purpose:

> The pursuit of information narrowly conceived—treating acquaintances as apps or
> search engines—forecloses on other levels of conversation. For instance, a friend
> may give inefficient directions compared with a map app, but their directions will
> also convey affect, memory, glimpses of how they understand the world, all aspects
> that could potentially reinforce or clarify the friendship.

Munger recognises the ability of algorithms to learn how to produce more
social forms of communication—perhaps through a suggestion that a birthday
is mentioned. Yet, if everyone knows that the mention of a birthday in an
email is likely the product of a smart-tech prompt and not human memory,
what does this mean? Smart technology can learn to be seen as thoughtful, but
human conventions surrounding thoughtfulness depend on the actual act of
remembering and the sense of care this conveys (Munger, 2021). Munger speculates about what happens when thoughtfulness becomes automated. Will we
become suspicious of personalised communication and come to think less of
it because of its lack of authenticity? Finally, Munger considers how humans
may creatively respond to the algorithmic framing of communication. He
suggests that in order to convey genuine forms of human thoughtfulness and
care (a sense that I know and remember very specific things about you, and this
shows that you matter to me), people may start to encode intentionality into
their written communications (Munger, 2021). In this context, Munger (2021)

argues that 'Linguistic novelty might create brief spaces of freedom, outside the machine's training data and thus be distinctively significant to humans'.

Munger's analysis of smart digital communication exemplifies the diagnostically specific analyses of human–smart-technology interfaces that we favour in this volume. This is a form of analysis that eschews more general techno-utopian or dystopian perspectives. It also blurs the boundaries between what we may consider human and technological (notice how smart technology may seek to become more human in its communication styles, just as humans become more machine-like in their speech). As with Munger, we explore the specific entanglements of humans and smart technology and consider the specific implications and ambiguities of these entanglements for different aspects of social life.

Human–Smart-Tech Interface Tropes

There is a series of existing tropes that can enable us to begin to make sense of the connections between humans and smart technologies. Some have suggested that the human–smart-technology interface is characterised by the technological production of digital humans. In his 2017 book *We are Data*, for example, Cheney-Lippold claims that smart technologies gather data about humans to construct digital selves. These digital selves are equivalent to our algorithmic reflections and are used by smart-tech companies to predict our likely needs and to generate personalised behavioural prompts (Cheney-Lippold, 2017). The smart-tech fabrication of digital humans is significant because it can serve to disrupt human identity and our ability to control who we are in the digital realm. According to Cheney-Lippold (2017), our digital gender identity or racial profile may not correspond with our actual self—it is merely what works best for algorithmic purposes. But, if this technological production of digital humans is inaccurate, the proprietorial nature of algorithms means that we often have limited power to change the virtual approximations of ourselves.

Sometimes, it appears that the human–smart-tech interface is characterised by practices of relational obscurement. In his pioneering analyses of roboticists, Berreby (2021a) explores how human input into smart technology is often hidden, while in other instances the technological aspects of smart devices are obscured beneath a human veneer. It appears that the downplaying of the role of humans in the programming, reprogramming, and maintenance of smart technology occurs because of a desire to both enhance the allure of technology's power and downplay any association with human fallibility (Berreby, 2021a). Taylor (2018) describes this process as *fauxtomation*—or the production of dehumanised accounts of the automation processes. The notion of fauxtomation is, of course, evident in the old joke about the smart factory: all a smart factory needs to run is one human and one dog—the human is needed to feed the dog, the dog to make sure the human does not go anywhere near the machines! But while fauxtomation may produce an enchanting sense of magic around smart tech, in its exclusion of associated human labour inputs it produces a very partial picture of the nature of smart technology. From Taylor's social feminist perspective, fauxtomation enables the devaluation of labour through the false assumption that it can be easily replaced by machines.

Berreby's (2021b) work also identifies a counter tendency within the relational obscurement of the smart-tech–human interface. This counter tendency is characterised by the anthropomorphising of smart technology. This process involves giving smart tech human characteristics. The anthropomorphising of smart technology is evident in the reassuring human voices of Alexa and Siri. It is manifest in the creation of robots that mimic the physical features of the human body. It is also evident in attempts to give smart-tech devices a sense of humour. Anthropomorphising smart tech appears to play an important role in the social acceptance of new forms of technology within our lives. Giving smart technology human features makes it easier for many to trust it. These forms of trust are particularly important as we develop ever more intimate relations with smart tech within our workplaces and homes.

A common account of the relations between humans and smart tech is one of human displacement by technology. While this account is connected to the discourses of relational obscurement, it is distinct. Human displacement by technology is not about hiding the human labour input that animates smart technology, it is about its actual replacement. In his analysis of the history of automation, Carr (2016: 19–41) reveals a long record of accounts of human displacement by technology. According to Carr, the narrative of the incremental replacement of humans by technology has both positive and negative discursive inflections. The positive discourses suggest that smart technologies will free humans from monotonous labour and enable us to enjoy more leisure time and to pursue more fulfilling and creative endeavours. The negative discourses are equally familiar, as they claim that smart technology will result in escalating unemployment and the loss of economic autonomy and purpose within human life (Carr, 2016: 20–34). Narratives of human displacement by smart machines may fail to recognise the obscured human labour (both paid and unpaid) that service the smart-tech world. Nevertheless, they do reveal one of the most important questions associated with smart technology: If smart tech's goal is to provide a substitute for human physical and cognitive labour (in increasingly sophisticated fields of human expertise), what are the consequences of this for what it means to be human (Carr, 2016: 18)? For Carr, it means that humans will increasingly become passengers in this smart-tech world (quite literally in relation to smart cars). As passengers, we will undoubtedly be liberated from many physical and mental tasks that are tedious and undesirable. But Carr suggests that the human displacement by technology has costs that go beyond the loss of income. According to Carr, our desire for labour substitution by machines is a classic case of *miswanting*, or humans not really knowing what is good for them. Work tasks and endeavours are important for human wellbeing, enthusing us with a sense of purpose and accomplishment (Carr, 2016: 14–18). It also turns out that humans are generally happier while at work compared to at leisure, as we have things to focus on outside of our own concerns (Carr, 2016: 14).

Related to the trope of human displacement by technology is the more troubling discourse of the technological domination of humans. While generally more unambiguously dystopic than the smart-tech narratives we have discussed in this section, notions of technological domination do come in different dystopian shades. First are those perspectives that suggest that through processes of digital surveillance and monitoring smart systems are able to understand and control human conduct in historically unprecedented ways (Zuboff, 2019). We will discuss Zuboff's theory of surveillance capitalism and its implications for how we interpret the smart-tech society in greater detail below. But at the centre of Zuboff's analysis is a suggestion that the rise of smart tech (and, in particular, the Big-Tech industry) is predicated on the exploitation of human nature itself. This exploitation process is connected to a loss of privacy and the freedoms that this brings, but also to a loss of control over our behavioural options and choices (see Chapter 4). The smart-tech domination of humans can, however, be seen beyond the realm of digital surveillance. It can also be the indirect product of simply working alongside smart machines. Some commentators have noticed that, far from liberating humans from physical and cognitive labour, smart technology can actually increase the demands placed on human labour. O'Connor (2021) reflects on the use of smart technology and automation in online retail warehouses (it is claimed that the market for automation in online store warehouses will increase from $15 billion in 2019 to $30 billion in 2026). According to O'Connor (2021), the creation of smart warehouses is not seeing technology entirely replacing human workers but rather warehouse jobs becoming 'part-human part-robot'. According to one warehouse manager, the reasons for this are the fact that they 'struggle to find the robot that will be able to handle a bag of plaster of Paris, a bit for a jackhammer, a galvanised steel garbage can, a saw blade, and a 5-gallon bucket of paint' (quoted in O'Connor, 2021).

O'Connor's analysis suggests that rather than easing the labour demands on humans, hybrid warehouses are increasing the intensity of the working environment for people (see Chapter 6, this volume). While smart robots may do the heavy lifting, humans are having to physically keep up with them and raise their own levels of productivity. According to O'Connor, this is not a form of technological domination that sees machines simply controlling humans, but rather results in humans becoming machine-like. To avoid the dehumanised intensification of work, O'Connor (2021) suggests,

> If we are to have robot colleagues, we need to design processes around the strengths and frailties of the humans, with ways for them to voice problems, propose solutions, and claim a share of the productivity gains. In other words, we must make sure the robots work for us, and not the other way around.

O'Connor's notion of robot colleagues provides a natural segue into the final trope of smart-tech–human relations: the notion of human–technological enhancement. Claims that smart technology can lead to various types of human enhancement take several forms. Their most extreme expression can, perhaps, be seen in the words of Tesla and SpaceX CEO Elon Musk when speaking at the World Government Summit in 2017:

> Humans must become cyborgs if they are to stay relevant in a future dominated by artificial intelligence ... There will be fewer and fewer jobs that a robot can't do better. If humans want to continue to add value to the economy, they must augment their capabilities through a merger of biological intelligence and machine intelligence. If we fail to do this, we'll risk becoming 'house cats' to artificial intelligence. (Musk, 2017, quoted in Rashid and Kenner, 2019: 20)

We discussed the figure of the cyborg in Chapter 1. As a fusion of cybernetic smart tech and organics, cyborgs can be conceived from a range of perspectives. In common myths cyborgs are an evolutionary threat to humanity, embodying the technological replacement of people by superior beings. But cyborgs can also be interpreted as a form of mutuality between humans and smart technology. Despite its hyperbole, it is this vision that informs Musk's reflections of an augmented merger between biological and machine intelligence.

*

We will return to the figure of the cyborg, and explore its wider epistemological implications for our analysis, later in this chapter. But as we conclude this section, we want to reflect in a little more detail on the issue of biotechnological mutuality. Contemporary research indicates that at this stage of technological development, optimal decisions and outcomes tend to emerge from partnerships between humans and smart technology. Such insights have led Pasquale (2020) to call for more humane systems of automation. In his book *New Laws of Robotics*, Pasquale argues that, 'A humane agenda for automation would prioritize innovations that complement workers in jobs that are, or ought to be, fulfilling vocations. It would substitute machines to do dangerous or degrading work ... This balanced stance will disappoint technophiles and technophobes' (2020: 4). We are sympathetic to notions of humane automation precisely because they cannot easily be reconciled with visions of techno-utopias and dystopias. They invite context-specific assessments of the types of human–smart-tech relations we desire. They also suggest that the form of smart-tech–human partnerships should not just be determined by their efficiency, but also on the basis of their impact on humanity. Humane automation thus calls not only for appropriate technological support for human life, but also the placing of limits on the smart-tech society to best preserve human autonomy, dignity, and purpose. In this context, humane automation is not just an analytical framework, it is also a normative position. In addition to seeking to facilitate the enhancement of the human condition through strategic partnerships with smart technology, it calls for greater human input into the form and design of the smart-tech society. We will explore the ideas of humane automations later in this chapter. At this point, however, we draw attention to the fact that narratives of human–technological enhancement do not necessarily have to involve an evolutionary celebration of an inevitable utopian fusion of smart technology and human biology (as Elon Musk would appear to suggest). Humans may see their capacities enhanced through partnerships with smart technology, but these partnerships do not have to be forced on humanity by Big-Tech. Humane automation reminds us that the smart-tech society can be a society within which human practical capacities are enhanced without their democratic rights, dignity, and autonomy being simultaneously stripped away. As a novel narrative trope of the smart-tech society itself, humane automation perspectives facilitate an appreciation of the ambiguity of human–smart-tech relations, while also suggesting that it is not too late to do something about the nature of those relations.

In the remainder of this chapter, we explore the conceptual frameworks that inform the tropes of human–smart-tech relations outlined above in greater depth. Each of these frameworks will, to some extent, inform the analysis of the empirical evidence we explore later in this volume. We also use this theoretical review to establish and justify the particular techno-social phenomenological perspective that defines our study of the smart-tech society.

PERSPECTIVE ON THE HUMAN–SMART-TECH INTERFACE #1—DIGITAL REALITIES AND DATA EPISTEMOLOGIES

The conversion of the world into digital form is a defining characteristic of the smart-tech society. The ability to be able to produce, monitor, share, and analyse real-time digital data from all around the world is something we tend to take for granted. This digital capacity is of course the product of a long and complex history going back to at least the late 1960s and the early iteration of what we now call the internet (then it was the ARPANET; Keen, 2015; Ball,

2020). The internet/ARPANET started life as a military-sponsored experiment designed to explore the potential for command and control in digital network communications (Ball, 2020: 30–31). Its core design, however, sought to 'let different machines connect, to scale, to work without central command and control' (Ball, 2020: 35). To put things another way, the early protocols which were developed to enable computers to communicate with each other were deliberately designed to facilitate a freedom of digital connection that would ultimately lead to the exponential growth of digital communication around the world. As Ball astutely observes, the early architects of the internet 'protected the nation [US] from one kind of technological surprise [the military-digital complex], only to create another' (the unregulated and perhaps unregulatable growth of the modern internet) (2020: 30). What now seems clear is that the design protocols of the internet enabled the geographical spread of smart technology to be achieved very quickly. In more recent times, the low costs of digital monitoring devices has also greatly expanded the reach of smart technology (Zuboff, 2019). Zuboff argues that the spread of the IoT is a fairly speculative activity: with the costs of monitoring devices being so low, there is little to stand in the way of embedding more digital monitoring devices into our world in case they become useful in the future and offer new revenue streams.

It is interesting, and perhaps troubling, to think that the digital foundations of the smart-tech society have been established on the basis of the ease of connection and cheapness of associated technology, rather than on a clear design purpose. But what is clear is that these foundations have facilitated an unprecedented ability to construct a digital world and to know the world digitally. There is a significant body of critical scholarship that explores the implications of digital knowledge and communication for human life. This literature does, of course, speak directly to the processes associated with the technological production of digital humans that we spoke about in the previous section. Much of this scholarship is beyond the scope of our specific concerns with smart technology. What we are primarily interested in in this section are the implications of digital forms of data gathering and communication for the feedback systems that connect humans with smart technology. In essence, we are interested in the consequences that derive from the fact that the human–smart-technology interface is a digital one. As we will see, the digital nature of the smart-tech world has implications for the forms of learning and self-awareness that are associated with smart technology (Figure 2.2). We also claim that the processes associated with digitisation have implications for perception, judgement, and questions of truth.

The primary impacts of digital modes of communication and monitoring on human knowledge and judgement are associated with what Beer terms data-informed knowledge (2019: 5). Data-informed knowledge is character-ised by the use of numerical data, statistical methods, and algorithmic analysis

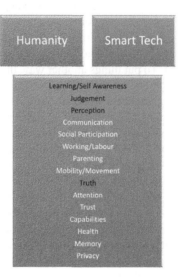

Figure 2.2   *Properties and processes affected at the interface of humans
                 and smart technology in relation to the technological
                 production of digital humans*

to produce understandings of the world. The emergence of digital life does
not axiomatically mean that knowledge sharing, and analysis, should be data
oriented: we all use digital communication to relay forms of knowledge that
evade easy quantification. Data-oriented knowledge does, however, enable the
maximisation of certain potentials that are associated with the contemporary
deluge of digital information. When presented as data, the signals of digital life
become easy to abstract from the particularities of daily life and become subject
to statistical comparison. When gathered at large enough scales, data-informed
knowledge can also be fed into algorithms, and patterns and inferences can be
quickly gleaned. When expressed as data, knowledge can also be quickly processed and fed back to interested parties (Kitchin and Fraser, 2020). The speed
associated with the processing and interpretation of digital data is central to the
operation of the smart-tech society. Acting on smart feedback is often dependent on that feedback being timely, which within the accelerating dynamics of
the digital world means it must be essentially instantaneous. So, data-informed
knowledge enables the functional optimisation of the digital world through the
production of information that can be processed at large comparative scales
and at high speeds. But there are costs and implications to this knowledge
abstraction model.

Critical studies of data-informed knowledge focus attention not just on what is to be gained from data-oriented imaginaries, but what the limitations and implications of these worldviews are. Beer's 2019 volume *The Data Gaze: Capitalism Power and Perception* personifies all that is best about this critical literature. According to Beer, the rise of data-informed knowledge is central to emerging norms concerning questions of authenticity, value, and truth. Beer deploys a loosely Foucauldian-informed notion of the data gaze as a framework to explore the epistemological, political, economic, and moral implications of seeing the world through the optics of data. What is, perhaps, of most interest within Beer's analysis is the way in which he explores the promises of the data gaze while also exposing the arbitrariness of its logics and practices. Beer's analysis of the promises of big data is undertaken with enduring scepticism. It is nevertheless insightful. Beer (2019) observes how the panoramic vistas facilitated by the data gaze promise 'insights into things that were previously outside of our field of vision.The result is that hidden value can be extracted from the depths of the structures in which we live' (26). We acknowledge the potential for data-informed knowledge to open up new perspectives on our individual and collective lives. These are macro and micro perspectives that inevitably evade our grasp, either because they operate at scales that are beyond our perception, or because they are obscured by their routinised presence in our collective unconscious. Our assertion is that the data on which the smart-tech society currently depends can expose novel views on reality and new truths. But, as Beer's analysis highlights, there are specific qualities to the contemporary data gaze that inhibit its capacity for human enlightenment and enhancement.

Central to Beer's theory of the data gaze is the partiality and bias of the data-informed knowledge that informs smart-tech systems. Data-informed knowledge is often celebrated on the basis of its unprecedented scale (see the discussion of Anderson, 2000, above): as something that supersedes human cognitive capacity to be able to see and interpret the world. According to Beer, however, the purported authenticity and precision of the data gaze is more a promotional tool than epistemological reality: 'The way in which data are seen is crucial to the power that they afford and the possibilities that are available for the expansion of data-led thinking, judgement, ordering and governance ... Being data informed is often presented as an inevitable force to yield to' (2019: 5). This data boosterism is a product of the big data and Big-Tech industries whose economic interests are firmly located in the universal uptake of data-informed knowledge. Beer exposes the operation of the data analytics industry (with its algorithms and data intermediaries) and reveals how its interest is not in truth production, but in the construction of the human world as data-based:

> The data analytics industry emphasises a particular vision of the social world so as to present data analytics as the only real solution. The power of data might well be located in what those data are used to reveal, but behind this power is an industry of activity working to spread those analytics and the optic horizons of the data gaze. (Beer, 2019: 15)

The data bias that Beer thus exposes is not just one of sensationalised content, click bait, and echo chambers, but a prejudice towards data itself. When you are in the data analytics industry it is important that every social problem is seen to have a solution that can only be solved through a data-based perspective on it.

The bias towards big data carries with it additional prejudices that can affect the perceptions and judgements of those that use them. A focus on data tends to result in a narrowing of the intellectual field of vision: all knowledge that is not easily converted into binary form is discounted as marginal and irrelevant. Data-informed knowledge does not necessarily have to produce less complicated visions of the world around it (indeed its scope and scale can generate much greater complexity of perspective). However, in their compelling analysis of the functioning of algorithms, Amoore (2020) suggests that proprietorial analysis of data within highly obscured algorithmic equations and codes unhelpfully simplifies complex human situations. This simplification means that smart tech will always struggle to replicate the nuance of human judgement that has been forged in the evolutionary field of actually existing social complexity over millennia. These are systems of human judgement that are not just predicated on what is true or false, but on what is just and unjust, fair and unfair, acceptable and unacceptable.

Perhaps the most significant limiting bias associated with contemporary data-informed knowledge is its association with speed. The speedy gathering and analysis of big data is not only required by the real-time feedback loops associated with smart technology. According to Beer speed is a necessary part of the projected power of big data:

> The insights produced by the data are seen to be a representation of the world as it unfolds—rather than being a reflective process of looking back. There is no delay or gap in the knowledge being produced, rather these data analytics are depicted as providing continuous encounters with the actually existing world as it is in that moment. (2019: 22)

Speed not only makes data-informed knowledge convenient, it also suggests that it is somehow a direct, unmediated replication of the overfolding social world. Speed carries with it a bias towards immediate adjudication and an inability to suspend judgement.

Beer's analysis ultimately demonstrates that despite its potential to support and enhance human knowledge, understanding, and judgement, digital data and the associated analytics industries do not provide a Royal Road to truth. Data-informed knowledge is also an expression of economic power. This power not only derives from the nature of data itself, but from the speed and scope of the analytics processes that surround it (Beer, 2019). The conceptual arrangement of data-informed knowledge is thus not neutral; its construction is instead informed by certain arbitrary principles, including speed, commercial value, what has happened in the past, and the simplification of complexity. Taken together, these epistemological features embody what Beer describes as a distinctive data rationality, which serves to replace existing regimes of expertise and supplement established systems of human knowledge and understanding (2019: 24).

Beer considers the specific connections between the data gaze and rationality. He argues that the notion of smart technology is central to the ideological promises of the big data industry, suggesting as it does an ability to learn and a latent intelligence within data analytics:

> The analytics are not passive, rather they are presented as being intelligent and active devices that are able to learn, adapt and develop in the insights that they produce ... A picture of powerful intelligence emerges. They respond to particular needs, learning what is required, whilst also learning from the data that is accumulating. (Beer, 2019: 29)

Here we can see the particular significance of smart technology within our data worlds. The notion of smart technology suggests that the datafication of existence not only offers humans new, panoramic perspectives on social reality— it also unlocks the promises of automatised learning. The compatibility of data with algorithms and machine learning mean that humans can, potentially, be taken out of the learning loop.[1] As Beer observes, this is when *dataism* and smart technology raise troubling questions concerning human agency. Enhancements in the scope of data gathering can undoubtedly augment human perception, judgement, and decision-making (core components of what we associate with agency). But when linked to smart systems they can also bypass the need for human input into the identification of optimal judgements. Although Beer acknowledges that humans are likely to still be required to decide what to do with the outputs of smart-data systems, this still

---

[1]   For a broader discussion of the impacts of smart technology and data analytics on learning (and the emergence of *learnification* systems), see Knox et al. (2020).

results in a heavily circumscribed form of agency. This, of course, is not to deny the possibility that data hackers, activists, self-quantifiers, and more open data systems can help to reassert human agency in our smart-tech world (see Kennedy et al., 2015).

PERSPECTIVE ON THE HUMAN–SMART-TECH INTERFACE #2—ROBOTS, CYBORGS, AND THE PHYSICAL INTERFACE

Robots and cyborgs have haunted the human imagination of the future for over a century. Beyond their invocation within science fiction, they offer important analytical vantage points from which to explore human–smart-tech relations. Robots and cyborgs are of particular interest within this volume because they express material manifestations of the smart-tech society. They serve to remind that within the smart-tech society the human experience of digital life has physical as well as virtual dimensions. A consideration of robots and cyborgs offers a counterpoint to our discussion in the previous section of the production of digital humans—it reflects smart tech's manifestation in the social world. Robots and cyborgs help to draw our attention to the physical interactions and encounters between humans and smart tech. Analyses of robots and cyborgs do respectively offer different types of insights for our analysis. We begin this section with a consideration of emerging work on the social implications of robots.

Robots are generally defined as programmable machines that are able to undertake various activities autonomously. Not all robots can be classified as smart technology and not all smart-technology devices are robots. In this volume we are primarily interested in robots that have an ability to learn from their interactions with humans and the physical world and are not merely pro-grammed to carry out repetitive tasks (however complex). We also interpret smart robots to be machines that are able to conduct fairly complex movements and thus have a more prominent presence in human life than static smart-tech devices. We would thus see an autonomous car or a smart vacuum cleaner as being a type of smart robot, but we would not include a smart fridge in our definition. For us it is the presence and movement of smart robots in the spaces of public and private social life and their resultant interactions with humans that generate their most interesting analytical implications.

Work on the social implications of smart robots draws attention to important psychological and ethical aspects of the smart-tech society. From a psycho-logical perspective it is interesting to note that while not all smart robots take human form, the aesthetic mimicry of humans is often seen as being important to their social acceptability. Pasquale argues that 'putting a friendly face' on robotic technology is important to their commercial success (2020: 8). Human mimicry is, of course, about more than physical form; it also relates to the ways machines move, their tone of speech, and their choice of language. Pasquale suggests that the forms of human mimicry evident in smart robots (and other smart-tech systems) can involve a pernicious 'counterfeiting of humanity' that can seek to establish human trust and acceptance through active deceit (2020: 8–9). In this context smart robots can deploy forms of relational obscurement wherein their technological form is obscured beneath a veil of humanness. Of course, the nature of smart technology means that forms of technological mimicry can be continuously and automatically refined to maximise human acceptability (although smart robots cannot effortlessly change their physical form).

It is important to acknowledge that developments in robotics are a long way away from hiding robots' presence within human society—a robot may put us at ease with its humanness, but we will still know it is a robot. Nevertheless, the presence of machines within social life, which are constantly learning to be more human and thus obscure their technological form, does raise interesting ethical questions. There seems to be a double deceit in the robotic counterfeiting of humanity identified by Pasquale: smart robots seek to learn about humanness, so that they can be socially acceptable, and thus be able to learn more about humans. This learning cycle is a common feature of the smart-tech society, but when it comes to robots, it takes a more visceral form. Perhaps the conclusion to be drawn from Pasquale's analysis is that a more ethical deployment of smart robots within the social world would involve a more overt signalling of the non-human nature of everyday machines. In this context, the technologically uncanny form of robots can serve as a useful prompt for humans to be aware of their interactions with smart technology and consider what they actually want to share and obscure from its field of vision.

The work of Berreby (2021b) demonstrates that the presence of robots in everyday life does not only raise ethical questions about the impacts of robots on humans. According to Berreby (2021b), the emerging presence of robots in social life raises intriguing questions about the human treatment of smart technology. Although it is possible to imagine people developing forms of meaningful attachment to purely digital forms of smart technology, the corporeal nature of robots generates novel ethical questions of attachment and care. There is plenty of evidence that demonstrates the human proclivity to develop feelings for a machine (particularly things such as cars and lawn mowers). It seems plausible that smart-tech robots are likely to inculcate enhanced forms of human–technological attachment. These attachments may derive from the human aesthetics of such devices. They may also be connected to the fact that smart-tech robots have become what they are because of sets of learning relations with their human owners/operators. The notion that a smart-tech robot may be uniquely formed through its relations with particular humans is likely to mean that humans view it less like a tool and more as an artefact of socio-technological interaction. For Berreby, the presence of human feelings towards robots raises questions of the appropriate regimes of care. In the context of Pasquale's concern over deceptive forms of human counterfeiting, it could be argued that smart-tech robots are being designed to exploit the caring instincts of humans to protect their commercial property. But Berreby's analysis suggests that there is a more complex set of ethical issues associated with the human treatment of robots.

Berreby reflects on the case of HitchBOT (Figure 2.3), a robot designed by a team of Canadian researchers to road-trip across North America through the kindness of strangers (Matisse, 2015). Having successfully travelled across large parts of Canada and Europe, HitchBOT met its demise near Philadelphia, where it crossed paths with a stranger whose kindness it could not rely on. HitchBOT was essentially an experiment designed to explore the extent to which robots could trust humans. HitchBOT's destruction leads Berreby to consider what an appropriate ethics of care towards robots might be. He acknowledges that there may be legitimate reasons why humans may choose to destroy robots—particularly in the context of active forms of dissent similar to those practised by the Luddites in the 19th century. But Berreby argues that we should not take the destruction of robots lightly. The harming of robots may constitute one of the ways in which humans can relate to smart technology, but it is a form of relation that may have costs for humans as well as smart technology. While Berreby stops short of suggesting that robots should have rights, he does argue that a human ethics of care towards robots could be based on what they mean to other people. He also argues that avoiding humans causing harm to robots is important because it avoids normalising violent relations between humans (in this context Berreby discusses the example of protests against sexbots that emphasise the normalising of harmful sexual relations among humans).

Ultimately Berreby suggests that a concern with questions of ethics and care towards (smart) robots is important precisely because it leads to engagement with the wider issues that are raised through the introduction of novel technologies into the world. According to Berreby, the best way of avoiding the harmful treatment of robots by humans is by ensuring that the forms of robots that are most likely to elicit violence do not enter the world. In this context, an ethics of care towards robots is important because it is best served by a broader form of connection with the complex field of politics and economics that generate those technologies in the first place. This form of engagement stands in stark contrast to the passive acceptance of new technologies as the inevitable outcome of socio-technical progress. The materiality of robots presents us with emotively significant reminders of the emerging smart-tech society we are a part of. Robots offer us material prompts to become more actively engaged

Figure 2.3   *HitchBOT*

in emerging technological processes. Berreby's (2021b: 16) analysis thus resonates with Pasquale's call for an anticipatory ethics of robotics. In anticipatory
form, a smart robotic ethics recognises that our main socio-technological
challenges may be in the future. But an anticipatory ethics also suggests that
it is only by engaging with related ethical issues in their embryonic form that
we can ensure that the values we desire are designed into smart technology
(2021b: 16). What the work of Pasquale and Berreby demonstrates is that the
intrusive and disruptive nature of robots means they can play an important role
in facilitating forms of values by design that ensure the development of a more
ethically oriented smart-tech world.

*

The figure of the cyborg offers related but different analytical opportunities
and challenges to the study of the smart-tech society. The cyborg is a 'fabricated hybrid of machine and organism' (Haraway, 2004: 14). As with robots,
cyborgs can draw attention to the physical presence of smart technology in
our lives. But while robots remain functionally separate to humans, the cyborg
reflects a more intimate fusion of technology and human biology. The precise
nature of the connection between technology and organism evident in cyborgs
can vary, from technology essentially taking a ride on the human body (in the

case of smartwatches or Google Glass), to being more functionally integrated (as with the use of smart telemetric implants, which can be used to monitor chronic health conditions). Not all cyborgs embody a fusion of smart technology with human biology though. Sometimes cyborgs are equated with the use of various prostheses or implants, which directly support the enhanced performance of the human body (such as the use of cochlear implants—a neuroprosthesis that can help tackle hearing loss). However, as a cybernetic organism the term cyborg is more accurately applied when used to refer to the learning and feedback systems associated with the smart technologies that are found on or in the human body.

In Chapter 1 we discussed how the figure of the cyborg has often been associated with dystopian visions of the technological future. In terms of the smart-tech society narratives we have outlined in this chapter, the figure of the cyborg can be associated with fears of human domination by technology and hopes for human–technological enhancement. Our interest in the cyborg relates to its ambiguous and disruptive potential when applied to human relationships with smart tech. The analytical value of the cyborg is, perhaps, captured most powerfully in Haraway's much-discussed 'Manifesto for Cyborgs'. Haraway's Manifesto is an essay that was first published in 1985 in the *Socialist Review*. According to Haraway, the integration of cybernetic communication systems and biotechnology facilitated the emergence of cyborgs as both material realities and imaginative resources. As a material reality, cyborgs embody a new evolutionary moment in the history of humanity, within which technology becomes not just an external tool, but a functional part of human biology. Writing in the mid-1980s, Haraway was aware that she was only catching the first glimpses of what cyborgs might actually become. For Haraway then, the cyborg was perhaps most significant as an imaginative resource. As an imaginative category, Haraway argued, cyborgs are significant because of their disruptive ontologies. Cyborgs are intellectually disruptive because they blur the boundaries between biology and technology, evolution and intelligent design, nature and culture. Haraway thus observes,

> High-tech culture challenges [these] dualisms in intriguing ways. It is not clear who makes and who is made in the relation between human and machine. It is not clear what is mind and what is body in machines that resolve into coding practices. In so far as we know ourselves in both formal discourse (e.g. biology) and in daily practice ... we find ourselves to be cyborgs, hybrids, mosaics, chimeras. Biological organisms have become biotic systems, communication devices like others. There is no fundamental, ontological separation in our formal knowledge of machine and organism, of technical and organic. (1987: 35)

Haraway's assertion that '[I]t is not clear who makes and who is made in the relation between human and machine' could have been written in response to the digital surveillance and learning and feedback systems that are synonymous with the smart-tech society. It is in this context that we argue that Haraway's (1987) 'Manifesto for Cyborgs' can offer important analytical perspectives for theory building in the smart-tech world.

We are conscious of the dangers of postmodern relativism that surround Haraway's cyborg perspective. Nevertheless, we claim that it can make very specific contributions to the critical analysis of the smart-tech society. First of all, in disrupting well-ordered divisions between humans and machines, Haraway's cyborg is epistemologically plural. It rejects narrow expert analysis of the smart-tech society and values a multiplicity of perspectives on newly emerging socio-technological relations. In addition to valuing diverse perspectives, Haraway's analysis also suggests that we are conscious of what smart technology can tell us about the world it is cocreating (with the proviso that there is more to know about the smart-tech society than smart tech can reveal). Ultimately, by collapsing the division between subject and object, the cyborg invites critically engaged analyses that are located within the field of analysis and not at some objective distance (Haraway, 1987: 9). Second, we argue that Haraway's analysis offers a context within which we can avoid dualistic understandings of human–smart-tech interfaces. In Haraway's 'Manifesto', the figure of the cyborg serves to undermine essentialised understandings of race, gender, and sexuality. In the time of the cyborg, race, gender, and sexuality cannot simply be seen as natural, but must be understood as constructed, and ultimately contestable, social categories. In our analysis of the smart-tech society the notion of the cyborg offers a lens through which we can begin to understand how technology is transforming various taken-for-granted aspects of what it is to be human (including self-determination, memory, and judgement, inter alia). Critically, however, and in partial distinction from Haraway, we are also interested in the inevitable ways in which technology becomes humanised within the learning loops of the smart-tech world (or what Taylor, 2018, describes as 'artificial artificial intelligence').

Haraway's 'Manifesto for Cyborgs' is an invitation to think in non-categorical forms. We do not necessarily ascribe to all of the postmodern implications that this non-categorical form of analysis leads to. We do, however, claim that Haraway's cyborg offers the kind of radical and analytically non-deterministic framework of analysis that is likely to be needed to make sense of the novelties and complexities of the emerging smart-tech world.

PERSPECTIVE ON THE HUMAN–SMART-TECH INTERFACE #3—SURVEILLANCE AND CONTROL

Theories of surveillance (and in particular *dataveillance*) and control have been central to the emergence of narratives concerning the domination of humans by digital technology. Indeed, the ability to observe and monitor human life as a pathway to controlling it has arguably been the dominant theoretical concern of smart-technology scholarship. Concerns over state-based digital surveillance have a long and complex history. Many of these concerns can be traced to the events of 2001. That year was witness to the continuing rapid growth of the internet and related forms of digital communication. It was also the year of the 9/11 terrorist attacks in the US. The year 2001 thus gave rise to a partnership between Big-Tech and national government wherein state intelligence agencies were given access to citizens' digital lives and, in return, Big-Tech companies were allowed to grow with limited regulation (van Dijck, 2014; Wood and Wright, 2015). To put things another way, the agencies who should have been regulating Big-Tech became organisationally dependent on Big-Tech companies (van Dijck, 2014: 203). To the extent that 2001 saw the normalisation of practices of mass digital surveillance and paved the way for the unrestrained expansion of Big-Tech monopolies, it can be thought of as the origin point of the smart-tech society.

The first wave of mass digital surveillance has been characterised by van Dijck (2014) as being a form of *dataveillance*. According to van Dijck, this phase of digital surveillance was based on a trade-off between privacy and the cost of digital platform services. The trade-off involved people exchanging their metadata for free digital services (van Dijck, 2014). This metadata was often unconsciously left by users but could be amalgamated into surveillance-based intelligence which could then be used by governmental authorities to identify potential security threats. According to van Dijck (2014), the practices of dataveillance were a threat to established political norms and practices within democracy affecting privacy and freedom of expression.

The early phases of dataveillance are part of the origin story of the smart-tech society. Our concerns in this book, however, are not with state-based covert surveillance but the more mundane practices of surveillance that people willingly submit themselves to in their engagement with smart-tech infrastructure. These forms of surveillance are geared towards relatively subtle attempts to control behaviour (see Chapter 4), as opposed to the more coercive interventions that often follow state surveillance. They are also focused much more on corporate profit-making than governmental security (although these considerations frequently align). Our primary focus in this volume is thus with the emergence of what Zuboff (2019) has termed surveillance capitalism. As its names suggest, surveillance capitalism relates to the building of regimes of economic accumulation around the practices of data monitoring established in the early years of the 21st century. Specifically, Zuboff defines surveillance capitalism as: 'A new economic order that claims human experience as free raw material for hidden commercial practices of extraction, prediction, and sales' (2019: vii).

Zuboff argues that the technological domination of humans takes a troubling turn under surveillance capitalism. No longer is it just about technology replacing human labour, but human existence being the raw material for new capitalist economies. Zuboff's theory of surveillance capitalism is distinct from early work on digital surveillance to the extent that it is primarily Marxist and not Foucauldian in its orientation. Foucauldian approaches to dataveillance emphasise the impact that the threat of surveillance has on human conduct: producing self-regulating conformist behaviours that support government desires for social stability. Zuboff's analysis focuses on the economic logics that inform the drive to data surveillance. Zuboff's work does, however, have implications for human behaviour and behavioural change (for more on this, see Chapter 4 of this volume). It is in the context of behavioural manipulation that we most clearly see that surveillance capitalism is predicated on smart technology. According to Zuboff, surveillance capitalism is built on a 'global architecture of behavioural manipulation' (2019: vii). This behavioural manipulation is not about humans self-regulating their behaviour in the shadow of surveillance. It is about the use of behavioural feedback, which is able to orchestrate behavioural responses through digital nudges and coercion. Foucauldian accounts of behavioural power align well with a neoliberal belief that in the absence of total knowledge the best hope of achieving stability and freedom is to allow people to self-regulate their behaviour. According to Zuboff, however, the behavioural interventions of surveillance capitalism are more controlling than Foucauldian accounts would suggest—involving what she terms instrumentarian power—because they are based on a more complete knowledge system and analytical certainty.

Ultimately, Zuboff brands surveillance capitalism a rogue mutation of capitalism that has created an historically unprecedented concentration of knowledge and power. She also claims that it reflects '[a]s significant a threat to human nature in the twenty-first century as industrial capitalism was to the natural world in the nineteenth century' (Zuboff, 2019: vii). We are sympathetic to Zuboff's critique of smart technology and the forms of society it is generating. We do, however, argue that her analysis is unable to deal with the radical ambiguities that are a key feature of the emerging technological world. In this context, surveillance capitalism provides little analytical space to consider the varied human experiences of smart tech, and the myriad socio-technological adaptations that are clearly part of the smart-tech society (see Monahan et al., 2010). Zuboff argues that surveillance capitalism draws our attention more to an economic 'logic in action' than to the technologies of digital surveillance (2019: 15). We agree that remaining conscious of the systems within which technologies operate is essential if we are to avoid narrow empiricism. Nevertheless, we argue that there are significant insights to be gained from a focus on the technological apparatus of the smart-tech society as it is experienced and negotiated by those living in that world.

Zuboff's analysis is so comprehensive and varied that it inevitably offers analytical insights that support our project. Prime among these is the notion of applied utopistics (Zuboff, 2019: 404–407). Zuboff argues that surveillance capitalism engages in a form of applied utopistics to the extent that while it promises us a bright future of human emancipation, it tells us very little about what that society will actually look like (see Chapter 1, this volume). Whereas traditional utopias offer us a vision of where we could go collectively, accord-ing to Zuboff, surveillance capitalists 'reverse the normal sequence of theory and practice. Their practices move ahead at high velocity in the absence of an explicit and contestable theory' (2019: 406).

Zuboff deploys the idea of applied utopistics as a critical concept to contest the essentially incontestable claims of surveillance capitalists. We claim, however, that the notion of applied utopistics, and the sense of indeterminacy it conveys, provides a productive framework within which to interpret the smart-tech society as it emerges at the interface of smart technology and human life. This is not to say that surveillance capitalists do not wield an unequal share of the power needed to shape this future, but to recognise that there is more at play here than merely the will of surveillance capitalists.

CONCLUSION

The theoretical perspectives we have explored in this chapter provide valuable critical perspectives on the smart-tech society. They also enable us to under-stand the nature and implications of the more popular narratives of human–smart-tech relations we outlined earlier in the chapter. It is not our intention to merge these analytical frameworks. Throughout this book we do return to each of the theoretical frameworks outlined above to offer context and to gain some analytical traction of the issues that we explore. The particular value of each of these perspectives lies, to a greater or lesser extent, in their ability to offer critical analyses of the impacts of smart technology without axiomatically descending into techno-utopian or dystopian thinking.

In this book we are genuinely interested in the apparent paradox whereby smart technology is clearly ubiquitous and popular, and yet is also the source of significant problems and concerns. To these ends, our approach is best char-acterised as a form of *socio-technological phenomenology*. To be absolutely clear, we are not phenomenologists, nor do we profess to be rigorous adherents to its practices. Phenomenology is itself a complex field of philosophical inquiry with various forms and branches. Our approach could be described as a phenomenologically inspired approach to human–smart-tech relations. Phenomenology is associated with an enduring interest in things. A devoted attention to things is related to an analytical commitment to avoiding theoreti-cal generalisations. Of course, phenomenology recognises that no matter how much attention we give to things, our appreciation of the physical world is inevitably constrained by the limits of our perception. In this context, we inter-pret phenomenology as a field of inquiry that explores '[t]he meaning things have in our conscious experience' (Smith, 2018). Throughout this volume we thus give particular attention to the ways in which smart technology is given meaning in the human experience of its various forms. We understand the human conscious experience of smart technology as taking a variety of forms, including physical sensations (the buzzing smartwatch), intrigue (perhaps in digital technology's predictions), fear (of surveillance), ease of being (gen-erated perhaps through convenience and time saving), or stress (due to the demands of social media). In the context of this perspective, we recognise that there is more to know about the smart-tech society than can be gleaned from human experience (it is unlikely that humans can directly sense the full scale of the nature of surveillance capitalism outlined by Zuboff, for example). We argue, however, that the human meaning given to smart technology, derived from our experiences of interacting with it, is likely to play an important role in determining the likely form of the smart-tech society. We also argue that theoretically informed analyses of smart technology (perhaps expressed in depictions of cyborgs, data selves, or dataveillance) can contribute to and transform the human experience of technology and themselves be explored at the phenomenological level.

What we value within a socio-technological phenomenology perspective is the emphasis that is placed on what the human experience of smart technology is. This is important for at least three reasons. First, it works against the temp-tation to align studies of smart technology with existing frameworks of analy-sis. Second, it recognises that human meaning, however misguided, is a critical factor in the emerging form of the smart-tech world. If the smart-tech world truly is an evolving dystopia, it matters if people don't experience it that way. Third, and from an existentialist perspective, a concern with the actual nature of the human experience of smart technology values human engagement with the smart-tech society. It serves to remind us that despite the dystopic rhetoric of surveillance capitalism, and the euphoric pronouncements of technological determinism, that human agency (stimulated by experience) still matters. This not only avoids letting humans off the hook in terms of collective responsibil-ity for the smart-tech world. It also promotes the direction of greater human attention to what smart technology is and what type of society it is supporting.

There are obvious critiques of a phenomenological perspective on human–smart-tech relations. More general critiques of phenomenology argue that it places too much emphasis on the role of the human as the steady eye in the centre of our experiential storm. We consistently try to avoid trying to see 'the human' as a fixed point of reference within our analysis of the smart-tech society. We thus remain intrigued by the impacts of smart tech in reshaping human perceptive capacities in the form of the cyborg (a form of digitally augmented phenomenology). We are also interested in human encounters with their data shadows and selves and the potentially disruptive impacts this has on how we see ourselves. Although the analyses we present in Chapter 3 (on smart-tech predictions), Chapter 5 (on the smart-tech body), and Chapter 8 (smart-tech disengagement) are heavily influenced by phenomenology, in other chapters we consider the emerging nature of smart technology from other perspectives (including the state, corporation, and city). Our approach can also be critiqued on the basis that much of the operation of the smart-tech society deliberately operates in a digital shadow world and seeks to target the collective unconscious. In this context, it is clear that not only will much of how the smart-tech world operates be beyond human perception, some of it will also be deliberately obfuscated. It is in precisely such circumstances that the theories we have introduced in this chapter can be drawn upon to support and supplement a phenomenological perspective.

Our broad application of a socio-technological phenomenology is neither orthodox nor doctrinaire. We deploy it as a helpful analytical and methodological system within which to explore the emerging ambiguities of the smart-tech society. We also emphasise that the application of a phenomenological perspective on the objects and operations of smart tech raises some intriguing possibilities. Smart technology forms a dynamic loop with humans through which both technology and humans (and their social relations) are transformed. This means that if phenomenology is about how things appear in human conscious experience, when it comes to smart technology, the phenomenon of study is itself changing the nature of consciousness and developing its own form of consciousness (not a digitally augmented human phenomenology of smart tech, but a smart-tech phenomenology of the human world). This raises the prospect of a dynamic form of phenomenological inquiry, which can help us to creatively explore the intricacies of the human–smart-tech interface.

3. Prediction, personalisation, and the data self

INTRODUCTION

#Scaryanticipation
'My echo dot seems to be aware of what clothes I am interested in'.
(SenseMaker Survey Narrative, 2019)

This is the first of a series of three chapters that focus on more intimate aspects of the smart-tech society. Of course, as we explained in the previous chapter, when it comes to the smart-tech society, the personal is always part of the collective. Nevertheless, we find it useful to consider smart technology from the perspective of the individual, family, and home. In this chapter we focus our attention on perhaps the most immediate and obvious way in which we experience smart technology—in the act of prediction.

We understand the act of prediction in obvious and less obvious ways. In the realm of the obvious, prediction can be understood simply as a kind of a forecast: an anticipation of what will come to pass, or what may be needed. The predictive aspects of smart technology can relate to human and non-human futures: smart tech can learn to use weather patterns to forecast rain, or monitor patterns of human behaviour to foretell of a traffic jam. In this chapter we are primarily interested in the predictions that smart technology can make with regard to human desire. Here predictions offer a window through which to explore how smart tech perceives the inner self and projects human futures. Given that the behavioural sciences suggest that humans are consistently plagued by irrational decision-making and poor judgements regarding longer-term need, the potential benefits of smart-tech predictions are significant (see Chapter 4).

From a less obvious point of view, prediction is also important because it offers a shadowy glimpse of our data self. Cheney-Lippold (2017) defines data selves as the composite identities that have been algorithmically constructed for us in the digital realm. This construction is informative because of what it reveals about us, and because it exposes blind spots within our digital identities. Our data selves are in part the product of all the information we share about ourselves with smart technology (including text searches, location information, voice tone, exercise patterns, inter alia). But they are also a product of what smart technology can infer about us, and what it needs to infer about us (Cheney-Lippold, 2011).

According to Cheney-Lippold (reflecting on the work of Nicholas Negroponte), our data selves are best thought of as algorithmic caricatures of humans. They essentially involve the conversion of human biology, psychology, physiology, and identity into programmable bits (Cheney-Lippold, 2017: 11). Our data selves, though, are neither able to be, or designed to become, mirror images of our analogue being. They are constrained by what smart technology can and cannot discern about the human condition (Cheney-Lippold, 2017, for example, reflects on the racial blind spots associated with facial recognition technology). They are also conditioned by what smart tech commercially and legally needs to know about us. In this context, Cheney-Lippold describes the algorithmic determination of fame utilised by Google. The line between being famous and a more ordinary human is not based upon an epistemological inquiry into the nature of fame but is the outcome of Google's need to know quickly and easily to whom it must apply European data protection laws, and to whom this doesn't apply (2017: xi–xiii). On these terms, the gender of our data self does not have to correlate with our actual gender to be of commercial value to smart tech. Neither do our data selves need to be consistent across smart-tech platforms. As Cheney-Lippold states, 'Google's, Quantcast's, and Alexa's interpretations of my data are necessarily contradictory because they each speak about me from their own, proprietary scripts. Each is ambivalent about who I am, interpreting me according to their individual algorithmic logics' (2017: 6). Our digital selves are thus more and less than our analogue selves. They are more, to the extent that they are composites of trends and lifestyle signals we may not even be aware of in our own behaviours. They are less, to the extent that they are inevitably biased and partial.

Due to their proprietorial nature, most of us are blissfully unaware of our digital identities. In the act of making predictions about our needs and desires, smart technology does however afford us partial glimpses at our digital alter egos (a kind of Plato's Cave for the smart-tech society). When predictions are wide of the mark, we tend to find comfort in the apparent failings of their underlying algorithms. When they are accurate, it is often assumed that they are part of a system of pernicious surveillance. We discuss these everyday responses and their attendant lay epistemologies in greater detail below. Within this chapter, however, we utilise smart-tech predictions, and human responses to them, primarily to gain insights into the emerging nature of the smart-tech world. In this context, we are in part indebted to the work of Amoore (2020). In her book *Cloud Ethics*, Amoore reflects on a debate between Ludwig Wittgenstein and Alan Turing (2020: 13). Unlike Turing, Wittgenstein did not see predictions as purely the outcome of numbers, but as a reflection of the specific grammar of mathematical assumptions. Amoore thus encourages us to see algorithms not as inevitable statistical outcomes, but as specific ways of seeing and interpreting the world. By exploring how individuals interpret and respond to algorithmic predictions, this chapter considers how the deeper workings of the smart-tech society manifest themselves in mundane everyday contexts: how smart-tech interpretations are interpreted. While focusing on this particular relation between individuals and smart tech, we realise that the wider grammatical logics of algorithmic ways of perceiving and relating to the world will not be revealed fully (we are inevitably confined to our digital caves). As one of the most common ways of experiencing the actions of smart tech, however, we claim there is value in considering the phenomenology of predictions. If nothing else, this perspective enables us to observe human responses to the predictive loops of smart technology. Given that we are likely to move into a more pervasive, and intrusive, era of smart-tech prediction in the not-too-distant future (see Zuboff, 2019: 199–204), understanding these responses may enable us to discern the likely implications and character of this predictive future. Ultimately, the research presented in this chapter reveals a human tendency of distrust towards accurate smart-tech prediction. While this may indicate lines of human resistance to the control of the future, it could also support the development of more subtle acts of smart-tech prediction. Much of the discussion of prediction in this chapter has implications for the anticipation and control of human behaviour within the smart-tech society. While we briefly reflect on the behavioural implications of digital prediction, the issue of behaviour forms the central focus of Chapter 4.

PREDICTION PRODUCTS AND THE AGE OF DIGITAL PERSONALISATION

#Listeningtomyconversations
'Adverts for new slippers when my current pair were full of holes. I hadn't even been searching for new slippers'.
(SenseMaker Survey Narrative, 2020)

Before considering the results of a study we have conducted into human responses to smart-tech predictions, it is important to position the acts of prediction within specific practical and economic contexts. Smart-tech predictions can take various forms. These range from the simple—Google offering predictive text to complete your search terms—to the more socially complex— Bumble recommending a dating partner. In her analysis of surveillance capitalism, Zuboff identifies two broad forms of prediction that are associated with smart technology, 'behavioural data ... are applied to product or service improvement, the rest are declared as a proprietary *behavioural surplus*, fed into advanced manufacturing processes known as "machine intelligence," and fabricated into *prediction products* that anticipate what you will do now, soon and later' (2019: 8, emphasis in original). In the first instance, prediction is a necessary factor of improved service provision. The ability to predict our need on the basis of previous behaviours is precisely what enables Google to be able to provide such rapidly available and personally relevant search results. Here prediction is connected to convenience, time saving, and ensuring the competitiveness of a smart-tech service. Related forms of prediction can also be seen when Facebook recommends friends, groups, and events that are relevant to us. In the second instance, prediction becomes part of a behavioural futures market, which Facebook, Google, and Amazon commoditise to facilitate secondary commercial activity. This is what Zuboff terms 'prediction products': or the predictive glimpses into the future that can be bought and sold by those who are interested in us buying things from them.

However prediction products are perceived, it is clear that they are playing an increasingly important role in a diverse set of social contexts. These span personal health needs (see, for example, the PatientsLikeMe platform), policing (see Amoore's analysis of Geofeedia and riot predictions), clothing preferences (StitchFix), and just about everything we need to buy (Amazon). While offering users convenience, smart-tech predictions are also providing unprecedented levels of personalisation. The goal of mass personalisation has been a commercial aspiration for some time. Following Fordist systems of mass production and consumption during the 20th century, personalisation has emerged as a new economic orthodoxy in the 21st century (see Deloitte, 2015). According to Deloitte (2015), consumers are willing to pay a 20 per cent premium for personalised goods and services. Additionally, more effective personalisation also promises economic efficiencies (and cost savings) to companies as well as likely increases in sales. Of course, not all forms of personalisation are predictive: it is possible to personalise a service (on smaller scales) by simply knowing rather than having to predict the nature of the consumer in question. Early attempts to generate enhanced forms of personalisation centred on the segmentation of the population (Deloitte, 2015). Segments represent fairly broad bandwidths of personalisation: 'older traditionalist', 'young radical', or 'green consumer'. They are inevitably limited by the amount of categorical variation they can capture and are static to the extent they do not allow for the ways in which individuals may circumstantially move between categories. The enhanced data-gathering and learning capacities of smart tech are enabling new forms of more dynamic and flexible personalisation.

Zuboff provides insights into the nature of the digital processes that are associated with enhanced regimes of personalisation. According to Zuboff, enhanced personalisation of digital services is driven by a prediction imperative. The prediction imperative pertains not only to the ability of smart-tech systems to provide highly personalised predictions, but also the need and desire to ever broaden the scope and depth of prediction practices (2019: 200). At its heart, the prediction imperative is driven by a question: 'what forms of surplus [in digital behavioural data] enable the fabrication of prediction products that most reliably foretell the future?' (Zuboff, 2019: 200). According to Zuboff, in the first wave of smart technology's predictive expansion, personal data gathering focused on online behaviours (include search histories, dwell times, group membership, etc.). But, to be more effective at predicting your future, smart tech is already imbricated in the second wave of the predictive imperative: a move from the virtual world into the real world. Zuboff outlines two iterative phases within the second wave of the smart-tech predictive imperative. The first is the extension of digital data-gathering capacities: 'Extension wants your bloodstream and your bed, your breakfast conversation, your commute, your run, your refrigerator, your parking space, your living room' (Zuboff, 2019: 201). On these terms, extension is connected with the data-gathering capacities of the IoT: your heart-rate monitor, your sleep app, digital voice assistant, and the smart car, watch, kitchen, and city.

The second iterative phase of the prediction imperative is the drive for depth. Zuboff described this drive for depth in the following terms: 'The idea here is that highly predictive, and therefore highly lucrative, behavioural surplus would be plumbed from intimate patterns of the self. These supply operations are aimed at your personality, moods, and emotions, your lies and vulnerabilities' (2019: 201). Predictive depth is enabled by the many devices that are associated with the IoT (although it is also supported by ubiquitous computing, which has the potential to capture facial expressions and posture). But it is distinctive to the extent that it does not just want to know our behavioural patterns, but how these are connected to our changing character.

In the pursuit of predictive depth, the smart-tech society has made quantum leaps from population segmentation. Here personalisation is not based upon a prediction of a generic social group I may belong to. It is not even just based on a more individualised understanding of myself in data form. No, it seeks to close the gap between experience and my data self to the extent that it understands the mercurial nature of my personality as it changes diurnally, seasonally, and according to social context, moment by moment. All of this suggests that the prediction imperative is not about the construction of my data self, but multiple data selves.

Within this chapter we recognise that the predictive imperative perceived by Zuboff has not come to pass in its fullest form. Nonetheless, we suggest that smart-tech prediction has entered a sufficiently advanced phase to be generating social reactions. While inspired by the insights of Zuboff, it is also important to emphasise that this chapter does not axiomatically adopt her techno-pessimism (see also Chapter 2). We recognise that enhanced forms of prediction, and the forms of predictive infrastructure on which they depend, raise a series of significant concerns (including privacy, human autonomy—we say more about this, and connections to the prediction of digital addiction, in Chapter 4). But we also appreciate the economic, social, and personal benefits of predictive personalisation. Significantly, enhanced forms of personalised prediction appear to be central to enabling people to engage with the data deluge associated with the smart-tech world, and the strains this puts on attention and cognitive capacity (see Chapter 8). In exploring people's reactions to smart-tech predictions we thus remain curious about the role of prediction as both a comfort and a concern, and how, as both a comfort and a concern, personalised predictions affect ongoing social engagements with smart technology.

A NOTE ON METHOD—SENSEMAKING IN COMPLEX SYSTEMS

Before exploring the results of our study into human responses to smart-tech predictions we want to reflect on the method of research and analysis we deployed. We do so because the method we have used could be considered unorthodox. While this section is, we believe, important to understanding the nature of the results that we have generated, it is possible to interpret the results without a deeper appreciation of the methodology. So, if you want to skip this section, we won't be offended.

The SenseMaker Philosophy and Method

Interpreting individuals' responses to smart-tech predictions presents research challenges that are both unique and more general in nature. In a general sense, asking people's opinions about something always raises research challenges. Attention must always be given to the way in which your questions are contextualised: positive and negative framing of questions can generate inherent bias in any survey. Asking people to retrospectively reflect on an event (like a smart-tech prediction) also raises questions about the reliability of participants' memories. When surveying people there is an inevitable need to structure their response options in such a way that the researcher simplifies the complexities of the real-life situations under scrutiny. Finally, surveys (particularly those that use binary Likert scales ranging from 'strongly agree' to 'strongly disagree') tend to promote an unthinking consistency in response from (either predominantly positive or negative) participants. Beyond these general challenges, asking people about their responses to smart-tech predictions generates more specific challenges. Many of us are now being inundated by smart-tech predictions, many of which go unnoticed. In surveys there is thus a tendency to reflect on predictions in quite a generic sense, or to choose those that are most memorable.

These challenges are inherent in any form of qualitative social science research design. In this chapter, however, we report on results from a survey methodology which we feel offers some creative solutions to these methodological problems. SenseMaker is a research methodology that has been developed by David Snowden. Snowden was responsible for establishing IBM's Global Services Knowledge and Differentiation Programme, and he has an ongoing interest in the importance of narrative and tacit knowledge in organisational research. Snowden's (2002) work focuses on the challenges of conducting research within complex and emergent systems. His SenseMaker methodology represents a particular adaptation of the broader field of research associated with sensemaking in organisational contexts. Sensemaking is a research approach that seeks to shift attention from rational and planned responses and reflections to hypothetical and/or theoretical issues, to a study of the meaning that individuals attach to events and actions (Weick et al., 2005). Narratives are particularly important in this context, as they embody the plausible stories and socially meaningful understandings that people use in order to make complex systems manageable (psychologically and practically). In this context, narratives are not mere fictions; they reflect encoded responses that people individually and collectively draw upon to navigate their way through uncertain systems (Weick et al., 2005). As such, they can offer telling insights into the character and nature of the field of actions within which humans find themselves.

As a particular, and original, approach to sensemaking, Snowden's SenseMaker method has a series of features that are worthy of reflection. Conceptually, SenseMaker focuses on informal and everyday stories to determine what is *knowable* and can ultimately then be known about emergent systems (Kurtz and Snowden, 2003). In this context, SenseMaker is predicated on a rejection of the assumption of ordered and predictable systems. Although SenseMaker is generally applied in organisational contexts, we feel that it is well suited to facilitating an informal exploration of emerging experiences of the smart-tech society. As a complex and emergent system, the nature and social responses to smart tech are varied and evolving. It is thus highly beneficial to be able to deploy a framework that has been designed to work with a dynamic field of action. A SenseMaker approach acknowledges that there will be aspects of a complex system that cannot be reliably perceived and interpreted. What it seeks to actively avoid, however, is 'relying on expert opinions based on historically stable patterns of meaning [which] will insufficiently prepare us to recognise and act upon unexpected patterns' (Kurtz and Snowden, 2003: 469). In this context, SenseMaker does not draw on expert opinion to try and make the unknowable knowable. This is not to say that engagements with experts are not valued, but only in the sense that it can make knowable things known. The danger with expert opinion resides in the fact that it can make dynamic systems seem predictable and self-determining, when in fact they are anything but. Interestingly, many in the smart-tech sector appear enthusiastic to promote the idea that the smart-tech society is a technological inevitability (Zuboff, 2019). We utilise SenseMaker in order to avoid this deterministic reading.

To try and avoid our research providing an easy fit with either techno-utopian or techno-dystopian visions of the digital world, we utilise SenseMaker to explore 'how people perceive and make sense of situations in order to make decisions' (Kurtz and Snowden, 2003: 470). In this context, SenseMaker very much supports the broader techno-phenomenological approach favoured in this book (see Chapter 2). As we discussed in Chapter 2, focusing on human experience does not mean we remain ignorant of those things that inevitably escape human perception. Nevertheless, priority is given to human experiences (and associated stories) because these are ultimately central to the ways in which emerging systems are interpreted, used, changed, and rejected. As Kurtz and Snowden (2003) indirectly remind us, a techno-phenomenological approach is advantageous not only because it provides insight into the proximate human experience of, and meaning given to, technological systems, but also because humans have the ability to sense and perceive (admittedly to limited extents) larger patterns of existence and meaning of which they are a part. Also, if complex systems are ultimately self-organising ones (this seems particularly apparent in the context of the Web 2.0 user content that feeds the smart-tech society), it would seem important to explore social processes at the self-organisational level. None of this, of course, means that extant theories of the smart-tech society, such as notions of surveillance capitalism or the Californian Ideology, have to be dismissed. The key point is that they should not be allowed to predetermine understandings of the lived experiences of smart tech.

Although a SenseMaker approach prioritises focusing on human experience, it does not suggest that human subjects provide a kind of Royal Road to access ontological reality. In this context, SenseMaker itself is not premised on a desire to somehow get closer to the true nature of reality (Kurtz and Snowden, 2003). Instead, it is concerned with how people make sense of, and give meaning to, the realities they experience. These meanings should not be mistaken for reflections of reality. Rather, these meanings, however irrational and false they may be, are seen as significant to the extent that they guide people's relations to the worlds they are in and play crucial roles in facilitating and inhibiting changes in those worlds. So,

SenseMaker's focus on individuals does not require assumed rationality on the part of the subjects who are being researched. Indeed, SenseMaker surveys are designed to capture forms of human irrationality and inconsistency in ways that many others methods are not. In a related sense, SenseMaker does not give intentional behavioural epistemology priority. While it is assumed that some reported behaviours can reveal longer-term intentionality, it is also recognised that many behaviours are more impromptu and automatic. As Kurtz and Snowden felicitously observe, 'not every blink should be seen as a wink' (2003: 463). This suspicion of intentionality appears important within studies of the smart-tech society, where the rapid digitisation of everyday life is often associated within the unthinking uptake of new technological habits under the guise of supposedly intentional pursuits of happiness. People doing things with smart technology should not be automatically interpreted as an indicator of people having conscious intensions that inform those behaviours.

Having reflected on the conceptual orientation of SenseMaker, we now turn to its practical manifestations. To foreground the meaning that people ascribe to systems and situations, SenseMaker research begins with the identification of a specific prompt. The prompt is generally designed to deliberately provoke reactions and to provide a specific and meaningful focus for reflection. In relation to our study of human reactions to smart-tech predictions, we used the following prompt:

> It is claimed that advances in smart tech mean that the social media platforms you use will *know you better than you know yourself*. Reflecting on a social media platform that you use on a regular basis, give an example of when it anticipated your needs in an unexpected way. (SenseMaker Survey Prompt 2019, emphasis added)

Provocations like these are often seen as problematic within social science research because they tend to draw attention towards memorable or exotic examples. We acknowledge that a prompt such as this may not offer representative insights into the mundane aspects of people's engagement with smart technology. It is, however, advantageous in other ways. First of all, the memorable event is likely to have heightened significance to the participant, and thus make it easy for them to reflect upon how they gave meaning to, and made sense of, the event. Second, within emergent systems, it is valuable to use more extreme events as experimental devices to consider how individuals may respond to trends that, while not prominent now, may become more common in the future.

SenseMaker participants are encouraged to write short narratives based on prompts. We will reflect on specific examples of such narratives later in the chapter (some also appear as reflective provocations at various points

throughout the chapter). What such narratives offer, though, are meaningful reference points for participants to refer back to as they complete the
survey. As meaningful reference points, such narratives serve two key purposes. First, they encourage participants to reflect on how their encounters
with smart technology made them feel and the meanings that were attached
to this. It is hoped that by drawing attention to specific and meaningful responses participants will be less inclined to offer post-hoc rationalisations
of related encounters or offer more polished accounts of their experiences
that present themselves in more desirable ways (although there can never
be a guarantee of this).

Using the narrative as the context for asking questions, the SenseMaker
survey then deploys innovative ways to gather responses. Given that
SenseMaker operates on the assumption of complex and unordered systems, many related surveys deploy triads, as opposed to linear scales, to
capture participant views. Figure 3.1 provides an example of the triad that
was used within the survey that is reported later in this chapter. Individual
triads are designed to focus on either positive or negative aspects of the
topic that is under consideration. This encourages participants to confront,
in turn, both the positive and negative aspects of an experience. The geo-
metric form of a triad also tends to encourage non-binary ways of thinking
about an issue—or to put this a different way, thinking about opinions and
events as the outcome of the complex interaction between competing and
emerging processes.

The nature of the design of triads has four analytical benefits. First, it
makes addressing each question an intellectual exercise in itself, thus
avoiding easy recurse to another positive or negative response in an
easy-to-complete linear question sequence. Second, because each triad has
its own internal logic, it is much more difficult for participants to offer
logically consistent responses across the survey. This makes it much easier
to discern the forms of inconsistent, but nonetheless meaningful, responses
people have to smart technology. Third, rather than expecting people to
adopt a definitive stance on an issue, it encourages people to feel (cognitive-
ly and emotionally) the gravitational pull of competing priorities, desires,
and objectives, and to thus recognise the compromised nature of real-world
decision-making. Finally, and perhaps most importantly, the nature of
SenseMaker triads tends to avoid the simple determination of statistically
dominant responses to issues. The spatial nature of the triad facilitates the
visualisation of different emerging responses to events and processes. These
varied groups can be interpreted as the emergent social responses that are
part and parcel of complex systems. Figure 3.1 displays the responses that
people gave as to why they use social media. There is clearly a tendency for
respondents to suggest that social media is primarily important to aspects of

social participation. But rather than merely seeing this as a definitive result, the SenseMaker approach encourages us to take other clusters of responses (perhaps those in the middle of the triad, or between "Learning" and "Social Participation") seriously.

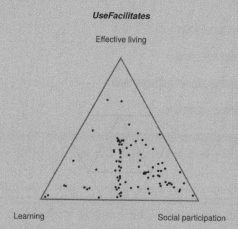

Figure 3.1   *SenseMaker survey triad—reasons for using social media platforms*

The SenseMaker Survey

The survey itself was completed in late 2019. It was shared as a digital link on various social media platforms, and through a series of email lists. While efforts were made to ensure that the survey was taken by people from a diverse range of backgrounds, there is inevitable bias in the survey sample. The survey was completed by exactly 100 people. As Figure 3.2 demonstrates, the majority of the participants were aged 18–24 years, with 72 per cent of respondents being 18–44 years of age. While these age biases cannot be ignored, they do, of course, reflect the fact that younger segments of the population are using smart tech much more than older groups (Ofcom, 2019).

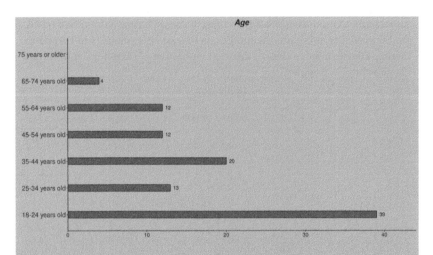

Figure 3.2   Age profile of SenseMaker survey respondents

The clearest bias generated by the sampling methodology concerned the educational attainment levels of participants. Figure 3.3 reveals that 68 per cent of those who completed the survey had attained a degree qualification or higher. Given that smart-tech users appear to be more evenly distributed among socio-economic groups than age groups, this is a more significant source of bias (Ofcom, 2019). However, when asked to rate their chosen SenseMaker narrative in terms of tone, a degree of balance was evident (Figure 3.4). Although 45 per cent of respondents described their chosen narrative as a negative one, 55 per cent indicated that their story was either neutral or positive. This would indicate, at the very least, that the survey was designed in such a way that did not frame social relations with smart technology as inherently negative.

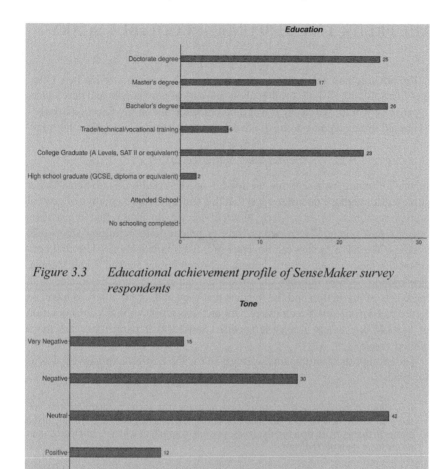

Figure 3.3   *Educational achievement profile of SenseMaker survey respondents*

Figure 3.4   *Narrative tone of SenseMaker story as determined by participants*

THE PREDICTIVE RESPONSE—'COOL, BUT SCARY'

#freakedmeout

'I'm not sure how relevant this is ... we have a Siri unit in our kitchen. One day I was talking [to] my brother about looking at new knives and new saucepans ... I would like to say it was a coincidence but on my Facebook page I started to receive promotional adverts on both items ... even with the right colours etc.'

(SenseMaker Narrative, 2019)

In the following two sections we reflect on the results of our SenseMaker survey. Although we do not suggest that this survey is representative of general human responses to smart-tech predictions, we do claim that it provides insights into potentially important sets of reactions to emerging smart-tech systems. We have broken our analysis into two main sections. The first considers the feelings that specific smart-tech predictions evoked within people. The second section outlines the meanings that participants associated with these acts of prediction and the impacts that they had, or are likely to have, on future patterns of smart-tech usage. Our analysis will combine a consideration of SenseMaker triads alongside specific SenseMaker narratives (like those quoted above).

The prompt that participants were given for the SenseMaker survey was as follows:

> It is claimed that advances in smart tech mean that the social media platforms you use will *know you better than you know yourself*. Reflecting on a social media platform that you use on a regular basis, give an example of when it anticipated your needs in an unexpected way.

A simple Word Cloud analysis of the vocabulary used within the participant narratives is informative. Figure 3.5 reveals the common use of phrases such as 'creepy', 'unwanted', 'scary', and 'listening'. While this would indicate the generally negative feelings that smart-tech prediction generated, a closer inspection of the narratives reveals a more complex picture.

The survey revealed that smart-tech predictions impact people's lives in both very mundane and dramatic ways. It appears that mundane predictions are often welcomed as a convenient but relatively incidental part of everyday lives:

> Facebook used my past search history to post sponsored ads based on recent history. I was able to find a good deal on a cute top. (SenseMaker Narrative 1, 2019)

Figure 3.5   *Word cloud representation of terms used within participant SenseMaker narratives*

I was looking for a gift for my other half for her birthday. I was a bit stuck, however Facebook had a sponsored ad on my timeline for a product which was ideal for them (it was a personalised book about a dog). (SenseMaker Narrative 2, 2019)

In these narratives, consumption predictions are seen as making people's lives a little bit easier and even facilitating some enhanced imaginative potential in relation to gift buying. Others reported much more dramatic impacts of smart-technology predictions. One participant revealed that smart technology had played a crucial role in helping him to contact his future wife:

#Metmywifethatway
'LinkedIn has a peculiar ability to know people that I do in fact know in real life, even if I haven't got their email or any other contact details'.
(SenseMaker Narrative 3, 2019)

In this context, LinkedIn's ability to predict desired connections with people who were actually known in real life provided a useful digital basis for establishing direct communication and long-term relationship building.

Digital predictions that support relationship building appear to be welcome (particularly when the prediction operates at a fairly generic level). Predictions generate much more ambiguous feelings, however, when they offer digital responses to other aspects of our private lives. Here is one poignant narrative reflecting on Twitter's response to a respondent's physical and mental ill health:

#Reading my notes

'I'd written on Twitter about being in hospital due to self-harm and feeling raw with the mental and physical pain involved. I was going to write further about needing to get away from the stress and needing a holiday—but before I could post that an ad appeared about taking a break!'

(SenseMaker Narrative 4, 2019)

We should perhaps not be surprised that Twitter's algorithm would interpret mental ill health as just another commercial signal for its personalised advertising. It is nonetheless problematic. It is not that the prediction was wrong—the participant was planning to take a holiday—although it was clearly reductionist. There does, however, seem to be something in the instantaneous, commercial response to this participant's post about mental illness that exposes the dehumanised nature of the prediction. The value of the prediction here is not the problem; it is its automatic timing, which serves to make it feel uncaring and insensitive. It appears that when predictions emanate from more socially sensitive and intimate aspects of the human experience (such as mental illness, or bereavement perhaps) that their nature becomes problematic. It is not necessarily that such predictions reveal something about our digital selves that we are unaware of. It is, however, in these more socially sensitive situations that the simplifying, commercial logics of smart technologies are more apparent. It is, perhaps, in these moments that we become conscious of the gap between smart tech and human empathy that algorithms have not, as yet, been able to close (Cooper, 2020).

Other feelings of unease were formed by predictive responses that smart tech generated in relation to personal relationships. One participant reflected:

#Breakuppickmeup

'Facebook—I went through a bad break-up and it suddenly started putting loads of inspirational quotes on my timeline—all I did was change my profile picture and delete my relationship status'.

(SenseMaker Narrative 6, 2019)

Again, we see that the evident distaste for the predictive response is born from mis-judged smart-tech infringements into personal distress. The participant openly shared news of their relationship break-up with Facebook. The predictive problem emerges from the apparent over-posting of inspirational messages. Feelings of annoyance towards the predictions here are not again because the prediction is inherently wrong, but because it embodies what is perceived to be a socially inappropriate (specially generic) reaction. It is interesting in this context to ponder whether if a friend had sent this participant an inspirational quote in the wake of their break-up it would have been perceived in the same way—probably not. When algorithmically generated though, generic quotes can feel less personally sensitive than when human selected (no matter how appropriate they are). They also do not reflect the act of caring that a human-sent post naturally conveys. Strangely, in this context it is possible that smart tech may actually have to be more socially sensitive than humans in order to be deemed acceptable in its responses to our changing circumstances. A smart-tech-selected quote may be much more relevant than a human-selected one, but here the issue is more about the medium than the message.

While predictions about personal matters appear to generate feelings of concern among smart-tech users, feelings of infringement are much stronger when users feel that aspects of their hidden, inner lives have been perceived by smart tech. One participant observed:

#Infringedupon
'It is difficult to call to mind exactly when, but at times when I've been feeling extra hormonal and maybe into getting pregnant and having a baby there has appeared a lot of sponsored articles about parenthood in my Facebook feed, despite me not having talked to anyone about this feeling'.
(SenseMaker Narrative 5, 2019)

In this instance the participant clearly feels that she has not shared information with Facebook about her inner hormonal state, and yet its parental predictions appear to have nonetheless discerned it. We will talk more about the connection between smart technology and the human body in Chapter 5, specifically in relation to bio-monitoring and the quantified self. What is clear at this point is that smart-tech predictions concerning our body and inner self are more likely to generate feelings of infringement. In this particular narrative, accurate predictions of undisclosed feelings generate a clear sense of transgression. The surprise that these predictions occurred 'despite … not having talked to anyone about them' is an interesting reaction in itself. We will speak more about everyday understandings of how smart-tech predictions work in the following section. But this narrative perhaps reveals a subject who is unaware of how demographic data and online activities could be used to generate accurate predictions of human biological states.

While accurate predictions can generate feelings of unease, transgression, and concern, inaccurate predictions generated different forms of frustrations. One respondent shared a story of digital prediction which was a rather striking example of digital misrepresentation:

#Dragqueenbetterthanreallife

'My Instagram feed is now full of drag queens—so is my explore page. I am
not a drag queen, nor am I any of the stereotypes that would put me into that
category. I'm a straight 20-something-year-old female who doesn't really like
wearing make-up or sparkly clothes, I just love what they do. Is that satisfying a need? Something so removed from my life?'

(SenseMaker Narrative 7, 2019)

In this instance there is a clear sense that a predicted interest in drag queens is
revealing a digital self that the participant does not identify with. Perhaps an
interest in certain subject positions, such as drag queens, skews Instagram's
algorithms in such a way that this becomes a dominant frame within which
to interpret the subject in question. However accurate or inaccurate they are,
Cheney-Lippold (2017) reminds us that our digital identity is primarily in the
service of Big Tech and not the subject it represents. If Instagram finds it commercially beneficial to assume we are a drag queen, this is likely to continue to
inform our digital identity, however far removed it is from reality.

An additional issue that was associated with digital predictions was distortion. Much has already been written on the ways in which smart technology
(and in particular social media) can generate reality distortion fields around the
individual (Lanier, 2018). The smart-tech filtering of content around individuals is driven by two primary factors: predictions about people's proclivities,
and controversial content. In the case of both the things we like and the polem-
ics we are drawn to, under surveillance capitalism, filtered content is selected
to maximise engagement (Osnos, 2018; Zuboff, 2019). One SenseMaker
narrative reflected on the impacts of such processes:

#algorithmtakeover

'Facebook—During the recent election [2019 UK general election] I received
lots of posts and ads relating to a single party—based on I presume my
viewing preferences/connections. This had led me to believe (wrongly) that
the majority of people had similar views and led me to presume a very different result to that which occurred'.

(SenseMaker Narrative 8, 2019)

Immersion in the algorithmically selected content we are likely to agree
with reflects a kind of virtual gathering of like-minded data selves. While
this digital clustering can generate feelings of belonging, reinforcement, and
support, it can also be discombobulating when individuals find reality does not
correspond to their echo chamber. This narrative is important, not because it
is surprising—of course it is not. But because it reveals that predictions rarely
operate in isolation (would you like to buy this car now?). They are part of
a much longer stream of filtered content that can easily create forms of epis-

temological distortion. It also reminds us that our data selves do not operate in isolation. They are constantly being algorithmically aligned with other data selves to drive continued forms of smart-tech engagement and network consolidation.

As Figure 3.5 reveals, the single most commonly expressed reaction people had towards smart-tech predictions was a feeling of them being 'creepy'. Creepy is an interesting adjective to choose to describe such predictions. It is suggestive of feelings of fear and unease with something that you don't understand. It is also indictive of the uneven division of learning identified by Zuboff (2019), whereby smart tech knows more about us than we know about it. Here are a series of selected SenseMaker narratives that all pertain to the theme of creepiness:

#Creepy
'When I ordered something through my husband's Amazon Prime, on one of his devices, to get free delivery my Facebook account (on my own device) seemed to know exactly what I had been up to and tailored my ads accordingly'.
(SenseMaker Narrative 9, 2019)

#Preemptive presumption
'Facebook ads are becoming more and more tailored. It started suggesting things that linked slightly with searches but now it's suggesting stuff I'd never even considered but feel like buying'.
(SenseMaker Narrative 10, 2019)

#Automaticrefill
'I was running out of a certain make-up product and advertisements came up for the product and offers for purchasing it. It was like it knew I was looking to purchase it again, without even searching for it'.
(SenseMaker Narrative 11, 2019)

#Spooky
'YouTube in the last year has become worryingly aware of my interests and own content. The suggested videos section on the main page or on the right-hand side includes some videos that are terrifyingly similar to what I have just watched and would like to watch further; sometimes on very peculiar topics that you'd think YouTube wouldn't be able to easily distinguish: certain musical pieces performed on particular musical instruments or in a particular style; mash-up meme videos—somehow the mechanism leaves out poor performances/videos or something that I wouldn't have liked'.
(SenseMaker Narrative 12, 2019)

In Narrative 9, a trans-network prediction (spanning a husband's Amazon account and the participant's Facebook account), the creepiness of smart tech appears to relate to the ways in which learning can jump across familial links and shared platforms. But in Narrative 9 we also see how predictions that transcend individual platforms provide insights to the user into the varied sources of data that are used to make up their digital selves. Narrative 10 is interesting because it reveals smart tech's seemingly uncanny ability to predict a subject's needs when they had never previously thought of buying certain goods. Of course, to smart technology, the data subject's direct awareness of a product is only one part of the broader algorithmic calculation of need. What is just as, if not more, important is what similar data subjects have purchased and could be used to drive new consumption in others. Narrative 11 is interesting because it reveals how feelings of suspicion towards smart tech can often emanate from digital predictions of analogue circumstances. The prediction of the need for a refill for a particular type of make-up suggests smart tech's ability to reach into the material fabric (or powder, perhaps) of everyday life to know not just what we need, but when we will need it. Here the predictive creepiness comes from the fact that smart tech is making predictions about non-digital phenomena (the non-smart make-up product). But in reality, smart tech knows nothing about the actual make-up container in the real world (only perhaps when it was purchased and how long it usually takes for someone of a certain demographic to buy another item). It just happens that in the real world the smart-tech prediction of purchasing need corresponds to the actual emptying of a make-up container. Smart tech doesn't know what it knows, but in this instance humans do, and that is deeply creepy to us. Interestingly in this context, it appears that Amazon wishes to move towards a more predictive mode of consumer supply, known as zero-click ordering. Goods will be sent to Amazon users when it is predicted they will be needed, and users can send any goods back that are not required (what is kept and what is returned will obviously help Amazon fine-tune its predictive apparatus; Galloway, 2017: 30).

Narrative 12 is interesting because it reflects on both YouTube and search predictions. Search predictions were not commonly reflected on in the SenseMaker survey we conducted; perhaps they tend to be associated with a more mundane type of anticipation of need (we will say more about this later in the chapter). The narrative is significant though due to the way in which it identifies the fine-grained predictive capacities of YouTube (Google), not only to predict which music videos the participant would like to watch, but also the appropriate styles, instrumental composition, and mash-up content. In this context, the creepiness of smart-tech predictions can derive from their ability to transcend platforms, operate at the interface of the real-world–digital divide, and anticipate needs we did not even know we could have. But, above all, the most reliable generator of feelings of creepiness is the accurate prediction. The more accurate the anticipation of need (perhaps Elio Pace's band's adapted performance of Billy Joel's 'We Didn't Start the Fire' for the 2015 England versus Australia Ashes series—well that's predictive musical gold for one of us, anyway!), the more suspicious people seem to be of it. As a human response to smart technology this seems important as it raises the critical question of trust. We will discuss responses to these feelings and their wider implications for the smart-tech society in greater detail in the following section.

There were two other feelings towards smart tech that our SenseMaker survey identified: annoyance and denial.

#Social media doesn't know me—I know me
'Instagram and Tumblr, and probably other social media like Facebook, have unique algorithms which assume what you like and want to see (more of). I've only just got Tumblr so we'll talk about Instagram. I follow 1946 people on Instagram and most of the posts posted by these people I don't ever see. Before Instagram changed its algorithm, it was in chronological order, but now your feed is organised for you. Now, what you see on your feed is not in chronological order and is customised by the site to show you what you like—it also does this with advertisements, and most of the time they aren't anything I'm interested in … It totally excludes the other 1900-odd people I follow from my feed. I hate it because I know what I like, I know myself, and I followed those 1.9k people because I was interested in them, but Instagram thinks it knows me better and chooses what I see anyway. It's clever, but also very annoying'.
(SenseMaker Narrative 13, 2019)

#Surprisinglydumbanddumberrecommedations
'Am surprised at how annoying the friend recommendations are ... I can see why the algorithm made the recommendation ... I am deeply p****d-off it used data from sources I wasn't expecting to be used and hadn't explicitly agreed to be used ... and actually I didn't want to be friends with the person … I just needed some intelligence about what my enemies were doing'.
(SenseMaker Narrative 14, 2019)

#Doesn'tapplytome
'To be honest, I can't think of a single example. The only social media platform I use is Facebook, and (for me at any rate) I would strongly contest the notion that it knows me better than I know myself'.
(SenseMaker Narrative 15, 2019)

Interestingly, Narrative 13 is one of the only accounts of prediction that considers the issues of social media feed ordering. The annoyance expressed in this narrative appears to derive from the sense of smart technology overriding the participant's desire to hear from his wider social network. The feeling of annoyance is articulated by the participant in relation to smart-tech's algorithms thinking they know better what the user wanted. In this instance at least, feelings of annoyance are connected to anthropomorphised depictions of smart tech as having a form of intellectual arrogance. In the case of Narrative 14, annoyance is a product of predictions clearly being made in an unsolicited way, and from sources of data that the participant did not feel they had consented to be used. It also appears to be connected to the participant using the platform in a way that smart tech had not anticipated (namely, to receive information about 'enemies' and not for getting friend requests). Narrative 15 is one of a relatively small number of responses that denied the notion that smart tech knew them better than they knew themselves. This perspective may be a product of the more limited use of social media by the participant that the narrative seems to imply. But feelings of denial are an interesting response to the rising influence of social media. Denial is, of course, a common response to perceived threats. In this context, however, denial appears to be about more than just securing cognitive protection from anxiety. Here, denial ('I would strongly contest') appears to reflect a deeper assertion of human nature's supremacy over technology. Such feelings have a long and complex history that can be discerned within religious ideology and even the political struggles of the working classes (see Carr, 2016: 19–41). Whether denial of this kind proves to be prophetic, or merely a futile rage against the machine, remains to be seen. The fact that smart technology's predictive potential can generate feelings of denial is the important thing to recognise here.

*

In the context of the varied feelings that the predictive acts of smart technology induce, it is informative to briefly consider the ways in which those who participated in the SenseMaker survey related to the social media platform they reflected upon. Figure 3.6 reveals that despite the feelings of unease associated with smart-tech predictions, there is a noticeable cluster of respondents who see personalisation as a primary benefit of using social media. Despite this desire for personalisation, participants tended to describe their use of their chosen social media platform as 'closed' and 'moderated' (Figure 3.7).

Even when asked to focus specifically on the positive feelings that they associate with smart-tech predictions, participants' responses tended to cluster around notions of 'surprise', with only a grudging nod towards being impressed (Figure 3.8). Notice also in Figure 3.8 how only one participant described feeling 'reassured' by smart technology's predictive prowess.

These results reflect what could be described as a predictive paradox at the heart of smart technology. One of the things that we value most about smart

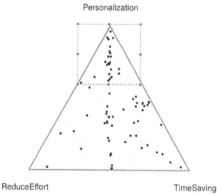

Source: SenseMaker Analysis Triad (2019).

Figure 3.6 *What benefits do you see in your social media platform knowing you well?*

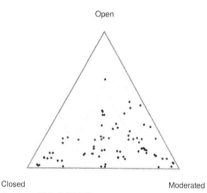

Source: SenseMaker Analysis Triad (2019).

Figure 3.7 *How would you describe your use of this social media platform?*

technology is its predictive potential. This is a predictive potential that both opens up unanticipated connections, experiences, and opportunities for learning, as well as makes our lives potentially easier. And yet, the more accurate smart-tech constructions of our data selves are, the more suspicious of its operation we become (particularly when those predictions appear to contravene social conventions of intimacy and appropriate response). This suspicion appears to lead to people moderating their relationship with smart tech more

Source: SenseMaker Analysis Triad (2019).

Figure 3.8   How did it make you feel when it anticipated your needs in this way?

carefully (thus, potentially undermining its ability to do one of the things we value most about it). The predictive paradox is perhaps summarised most effectively in the narrative quote with which we commenced this section: smart tech is 'cool, but scary'. In the next section we explore this tension between being cool but scary in greater detail. We consider the interpretative meanings that derive from people's feelings towards smart-tech predictions: or to put things another way, the lay interpretations that inform how people understand the operations of smart tech's predictive apparatus. We then consider the impacts that these feelings and meanings have for people's use of smart technology. While we do not propose to resolve the predictive paradox of smart tech, we do speculate on its implications for the future of the smart-tech society.

INTERPRETATIONS AND RESPONSES TO SMART-TECH PREDICTIONS

Interpretations of Predictive Power

Drawing on SenseMaker responses this section considers the kinds of lay interpretations that are used to make sense of, and give meaning to, the predictive power of smart technology. Given the opaque nature of the predictive processes associated with smart technology, these everyday interpretations provide valuable insights into the colloquial theories that emerge to fill the explanatory void.

Some of the SenseMaker narratives we collected reflected directly on inter-pretations of smart-tech predictions. In Narrative 16, one participant draws on a long-established theory of marketing in stating that smart tech does not so much predict need as creates it:

#stopsellingmeshitidontneed
'I would say that social media not so much anticipates needs as it *creates* them. The most frustratingly useful example of this for me would be that I am given advertisements for products if I have mentioned something that is of interest to me on WhatsApp'.
(SenseMaker Narrative 16, 2019)

The idea that predictions are less algorithmic discernments and more the product of repeat advertising and marketing reinforcement is a significant perspective. It is significant for at least two reasons. First, it may, in part, be true. In our concern with the novelties of Big Tech, it is perhaps easy to forget that conventional marketing strategies are still present in the smart-tech world (even if their power is greatly enhanced by digital personalisation and reach). Our focus on Silicon Valley should not mean we forget about the established powers of Madison Avenue. Second, it is significant because it reveals that in the absence of other more compelling explanations, it is common for humans to rely on established ways of understanding the world even if their explana-tory power has been compromised. The use of older explanatory frameworks can be reassuring and give us some semblance of perceptive control over novel developments. But reliance on older interpretive frameworks can be dangerous to the extent that it makes it easy to avoid inquiring more deeply into what is actually novel in the smart-tech society.

Other narratives displayed more detailed understandings of how smart-tech predictions are generated:

#creepy
'I use Facebook regularly, and, like many sites, Facebook uses custom ads and suggestions for different products and pages. You only have to search something once and it will suggest your search to you as an ad when you next log in. Once it suggested to me as an ad a drink that I was drinking at that very moment although I had never searched for it or typed the brand online. I assume (and hope!) Facebook just assumed I would be interested based on my other searches and likes and it was a coincidence'.
(SenseMaker Narrative 17, 2019)

This narrative displays an awareness of the role of personal search history in informing predictive responses. The reality of drinking an unsearched-for drink at the moment that Facebook predicts you would like this drink clearly gave this respondent pause. While this narrative displays an appreciation of how such an uncanny prediction could still be derived from search histories and likes, it appears to seek comfort in the idea of coincidence. Perhaps this was a case of coincidence, but the narrative raises the question of when accurate predictions are no longer seen as innocent coincidences.

A common explanation of the predictive power of social media platforms did however suggest a more sinister explanation. A series of narratives reflected on the potential of always-on smart tech to be listening in to people's conversations. The following narratives all made reference to the connection between prediction and smart tech's listening capabilities:

#Listeningtomyconversations
'Adverts for new slippers when my current pair were full of holes. I hadn't even been searching for new slippers'.
(SenseMaker Narrative 18, 2019)

'While planning for a trip to Austria with my friend on WhatsApp I noticed that I started to get adverts for Austrian winter vacations while completing Duolingo sessions. While I thought this may be a coincidence, I was a little worried given that WhatsApp is supposed to be end-to-end encrypted. Perhaps this meant that the encryption only applied to third parties'.
(SenseMaker Narrative 19, 2019)

#googleislistening #googleears #whatprivacy
'I often find that when I've been talking about something, and then I go to Google it or something related to it, the thing I've been discussing auto-fills as I type (even when it's something that you would expect not to be a very common search). Google is frighteningly good at predicting what you will search for. I find other sites or apps are less "intuitive" in that way'.
(SenseMaker Narrative 20, 2019)

It is well established that smart-tech companies have the ability to listen in to conversation, particularly through talk-activated systems, such as Alexa and Siri (Hern, 2019). It seems less likely, at least at the moment, that voice recognition software is the primary source of smart-tech's predictive accuracy. What is interesting about the recourse to a more sensationalist explanation of the predictive prowess of smart tech, is that: (1) it provides an easily understandable and well-established framework for interpreting a novel phenomenon; and (2) it tends to create an explanation of smart-tech prediction that actually obscures the true source of its power. As with suggestions that smart tech's power is just a rebooted version of *Mad Men*, or an innocent coincidence, listening-in conspiracies rely on easily understood human stories to interpret far more complex realities. By offering a more dramatised explanation of predictions, audio-spying theories also fail to grasp the more frightening human reality: smart-tech algorithms already have enough information on our data selves, and the data selves of others, to be able to make predictions that are much more reliable that those based on the rather laborious task of listening in to our everyday ramblings. This may seem dismissive of such explanatory narratives. Perhaps it is. But it does not have to been seen in this way. The reason that we utilised a SenseMaker analysis was to elicit meaningful feelings towards, and narrative explanations of, smart tech because they matter. They thus indicate how we as humans are making sense of this digital/evolutionary change in our existence. These stories are actively shaping our continued use of smart tech, and perhaps foreclosing other ways of being in the smart-tech world. Such explanations, thus, don't have to be true in order to matter.

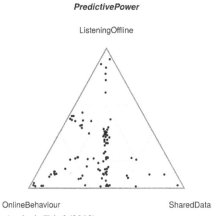

Source: SenseMaker Analysis Triad (2019).

Figure 3.9   *What do you attribute the predictive power of social media platforms to?*

Looking at participants' interpretations of the source of social media's predictive powers as a whole presents a more complex picture (Figure 3.9). Those who directly attribute it to listening offline are relatively few. The two most significant clusters are those that associate accurate digital predictions with online behaviours and the sharing of digital data, and those who see prediction as an outcome of online behaviours and offline listening (Figures 3.10 and 3.11). It is difficult to know whether the mixed clusters (Figure 3.11) are merely uncertain about the source of smart-tech predictions, or whether they see it as the outcome of off- and online surveillance. What is interesting about this cluster is that it perhaps offers the most accurate interpretation of the future of smart-tech operations. As Zuboff (2019) has revealed, if the history of surveillance capitalism has been based on web-crawling, the future will focus much more on life-crawling. The smart-tech society is now at least as much about the internet searching us (in all aspects of our lives) as us searching it.

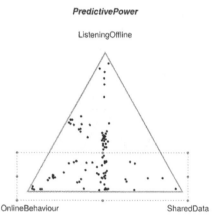

Source: SenseMaker Analysis Triad (2019).

Figure 3.10   *What do you attribute the predictive power of social media platforms to? The online cluster*

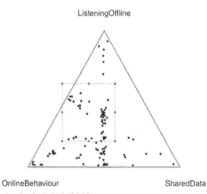

Source: SenseMaker Analysis Triad (2019).

Figure 3.11   *What do you attribute the predictive power of social media platforms to? The mixed cluster*

Responses to Predictive Power: The Consequences of the Predictive Paradox

In this final section we consider the impacts that personal predictions have on people's relationships with smart technology. As part of the SenseMaker survey we asked participants to reflect on the positive and negative responses and actions that smart-tech predictions generated. When asked for the positive actions that predictive power generated, only 32 per cent provided a response (see Figure 3.12). This would initially indicate that participants found it difficult to think of positive actions that derived from social media's predictions about them. None of those that responded suggested that social media's knowledge about themselves generated a greater sense of trust. This in itself is interesting given the general connection between knowledge and trust in human relations. In our social networks, getting to know someone well is generally a basis for trust building and the sharing of additional personal insights. Perhaps unsurprisingly, this relationship between trust and knowledge sharing appears to form an inverse correlation in terms of human relations with smart technology.

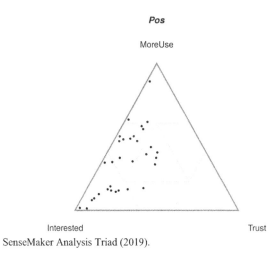

Source: SenseMaker Analysis Triad (2019).

Figure 3.12   The more a social media platform knows about me ...

When asked about the negative implications of smart-technology predictions, the number of responses was much higher (Figure 3.13). There is a clear cluster of responses, which indicates that accurate predictions generated feelings of distrust, concern, and an inclination to use the technology less.

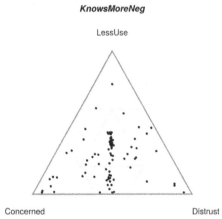

Source: SenseMaker Analysis Triad (2019).

Figure 3.13   *The more a social media platform is able to accurately*
                 predict my needs ...

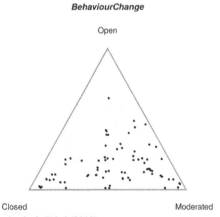

Source: SenseMaker Analysis Triad (2019).

Figure 3.14   *Behavioural response to accurate social media predictions*

Given the previously discussed uneasy feelings that smart technology generates in people, it should come as no surprise that smart-tech predictions
result in people trusting and using related platforms less. This is the predictive
paradox that we discussed in the previous section. Our results indicate that
the predictive paradox manifests itself in two distinct ways. First are actions
that seek to more carefully moderate or close off the sharing of personal data

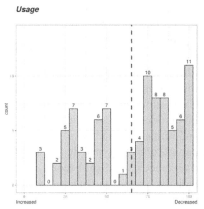

Source: SenseMaker Analysis Triad (2019).

Figure 3.15   *Change in usage over time following experience of social media predictions*

with smart-tech platforms (Figure 3.14). Second, is the tendency to start to use smart technology less. Figure 3.15 reveals the impact that predictive events had on participants' use of social media platforms. It is clear that, despite misgivings about smart tech's predictions, many people continue to use related technologies more. However, the majority in our survey indicated that, follow-ing the experience of social media predictions, they used the platforms far less. We speak in more detail about the challenges associated with disengaging with smart tech in Chapter 8.

CONCLUSION

In this chapter we have considered human reactions to the predictive acts of smart technology. In doing so we have considered smart-tech predictions as windows on what Cheney-Lippold (2017) describes as our data selves. While prediction provides a valuable phenomenological context in and through which to explore human feelings towards technology, it is also central to the underlying economic logic of the smart-tech society. According to Zuboff (2019), the whole edifice of surveillance capitalism is driven by prediction products and the prediction imperatives. A consideration of human responses to predictions thus provides a valuable hermeneutic perspective on both the intimate and systematic dimensions of the smart-tech society.

The central insight of this chapter has been to articulate what we have termed the *prediction paradox*. This paradox rests upon the fact that, while users value the accurate predictions that smart tech can make about their varied needs, the more accurate the predictions become about their desires, the more concerned the user becomes about the power and influence of smart technology. Cheney-Lippold's work reveals the partial, commercially driven inaccuracies that define our data selves. This chapter argues that users are also concerned when they experience predictions that appear to be based on incredibly accurate constructions of their data selves. Concerns over accurate predictions are particularly common when these predictions pertain to highly personal biological and emotional states. The paradox thus emerges that as smart tech improves and refines its predictive products, we are less inclined to share our digital data. In Zuboff's terms, this means that the better the predictive products of surveillance capitalism become, the more likely it is that the imperatives for predictive extension (into more and more aspects of our on- and offline lives) will meet social resistance.

It is our claim that this predictive paradox marks one of the key zones of negotiation that will define the future evolution of the smart-tech society. In Chapter 8 we explore in greater depth the different strategies that people are using to disconnect themselves from the smart-tech society. But what most interests us now (and in the following chapter), is what the prediction paradox means for the smart-tech sector. We can, of course, only speculate about this. However, if the prediction paradox is a generalised phenomenon, then it seems likely that it will be detected by the algorithms that undergird our smart-tech existence. If it is detected it will be seen as a cause of lowered smart-tech engagement, which undermines the prediction imperative and must be acted upon. But how can any system seek to compromise the very thing upon which it appears to depend—in this instance, accurate predictions? We would speculate that the experimental algorithmic response to the prediction paradox would involve the partial obscuring of predictive prowess. This obscuring process may involve deliberately feeding users a necessary dose of inaccurate predictions alongside accurate ones. It could also draw upon the behavioural strategy of partial reinforcement, whereby desired user behaviours are established and predictive encouragements are rationed so as to avoid overuse. Interestingly, this algorithmic practice of obscuring is a mirror image of the strategies of obfuscation that many users deploy to avoid the digital gaze of smart technology (Kitchin and Fraser, 2020). Precisely what the outcome of this interplay between obscuring and obfuscation will be is difficult to know, but it is clearly already underway. What is also clear is that these trends are likely to lead to a less transparent smart-tech society. As we will explore in the next chapter, this lack of transparency may suit the behavioural goals of the smart-tech sector.

4.   Behaviour and freedom

INTRODUCTION

This chapter explores the relationship between smart technology and human behaviour. In many ways the notion of human behaviour is easy to grasp. Behaviour is how we express ourselves, pursue our needs, and meet desires through action. Although we may all have an intuitive grasp of what behaviour is, developing a more precise definition is difficult. What, for example, is the difference between a behaviour and an action, or between a practice and behaviour? Is a basic action a form of behaviour, or should we consider behaviours to be more complex and longer lasting? Three things do appear to be clear about behaviours. First, the term is used to cover a broad range of varied human activities, including smoking, eating, exercising, driving a car, reading, working, meditating, and socialising, inter alia. Second, a belief in our ability to choose and control our behaviours (however misconceived) is central to our sense of what it is to be human. To this extent, humanness finds practical expression in our behaviours, and in how they differ from those of other species of things and animals. Third, the monitoring and modification of human behaviour is a central operating principle of the smart-tech society.

The term behaviour originates from the verb to have (as in be*have*). It refers to, consistently acting or operating in a specified way. Behaviours are generally associated with the actions of individuals and stem from the interaction of cognitive processes, emotional responses, and the surrounding environment. It is important to acknowledge that to understand human action through the notion of behaviour is immediately limiting. As we discuss in the following section, behaviour can suggest a simplified and psychologised understanding of human motivation. In these terms, human action can be seen as the outcome of individual minds or psyches. This perspective, however, neglects the complex contextual factors that shape human conduct over longer periods of time. Understanding human action as behavioural can also quickly lead to behaviouralist understandings of the human condition (Chomsky, 1971). Behaviouralists equate human conduct with the conditioned behaviours that are found in all animals, and that are determined by stimulus, response, and reinforcement. We use behaviour as a shorthand for human conduct because of the tendency within the smart-tech world to understand human action through behaviouralist frameworks. Thus, in addition to considering the potential impacts of smart technology on human behaviour, we also consider the practical and epistemological consequences of seeking to modify human conduct through behavioural strategies. As we will see, it is no coincidence that the smart-tech society is drawing on behaviourally informed understandings of human action to both justify and enact social change. In this chapter we explore what the likely implications of this behaviouralist strategy are for being human in the smart-tech society.

This chapter builds directly on the discussion of prediction in Chapter 3. According to Zuboff (2019), the very purpose of smart-tech prediction is ultimately to produce behavioural actuation. As we will see, predicting a human 'need' is often not enough to ensure that the need is actively pursued. In this context, behavioural modification is the fulfilment of the prediction imperative: the moment when the data self is used to manifest change in the real self. As we go through this chapter, we will see that smart-tech behavioural modification is often pursued at an unconscious level. In this context, we do not offer a phenomenologically oriented perspective on the behavioural aspects of smart tech (this kind of perspective is, however, offered in Chapter 5, through an auto-ethnography of wearable smart tech). Instead, this chapter reflects on existing analyses of digital behavioural manipulation in order to consider its longer-term implications.

THE BEHAVIOURAL IMPLICATIONS OF SMART TECH IN CONTEXT

The Great Recession and the Death of Anthropological Monsters

Before exploring the connections between smart tech and human behaviour, it is important to acknowledge a key shift in the way in which human behaviour is now understood by many. For much of the 20th century the social sciences have relied on a very specific model of human behaviour. This model is based on the figure of *homo economicus*. *Homo economicus* (or economic human) is a generalised account of human behaviour that relies on an assumption of rational action (Cohen, 2014).[1] *Homo economicus*'s ability to consistently behave in rational ways is predicated on four assumptions concerning human decision-making:

1.   That humans always act to serve their own best interest;

---

[1]   Even when psychiatry revealed drivers of our behaviour that are beyond our direct perception and control, psychology appeals to greater self-knowledge and improvement ultimately supporting the notion of *homo economicus* (see Rose, 1998).

2. That humans are able to make decisions in relative social isolation;
3. That when making decisions humans have access to all the relevant information; and
4. That people have the necessary time to effectively deliberate on their best course of action.

Sounds too good to be true, no? Well, it is, of course. *Homo economicus* is in fact such a gross simplification of the human condition that the French sociologist Pierre Bourdieu described it as an 'anthropological monster' (Cohen, 2014). In truth, *homo economicus* was never intended to be an accurate depiction of the human condition. Rather, it was more an approximation of a general tendency within human decision-making, which when taken in aggregate would provide a reasonable approximation of how large groups of people would tend to act over long periods of time.

Whatever its limitations, during the 20th century the figure of *homo economicus* gained intellectual power and influence. This power derived from its moral undercurrent. One of the early inspirations for the figure of *homo economicus* was Daniel Defoe's Robinson Crusoe. This account of a rugged, desert-island-based decision-maker suggested a certain moral character and entrepreneurship, which resonated strongly with libertarian visions of human nature. *Homo economicus* also supported certain political and economic projects. For the neoliberals of the late 20th century, the self-interested rationality of *homo economicus* was a central component of a belief system that suggested that unregulated markets could provide Western societies with both freedom and stability in the post-World War world (Whitehead et al., 2017). It was in these contexts that, despite the protests of many working in the behavioural and social sciences, the figure of *homo economicus* became a dominant model for understanding human behaviour.

Things started to change in 2008. In the Western world at least, 2008 will always be synonymous with the Credit Crunch, the Great Recession, and the ensuing era of austerity. But beneath the macroeconomic crisis of neoliberalism, was a more subtle epistemological crisis in how human behaviour should be understood (Jones et al., 2013; Whitehead et al., 2017). The reason for this great re-think derives not only from the unstable sub-prime markets of the US, but also from the newly available models of human behaviour that were becoming popular at the time. In 2009, the hardback addition of George Akerlof and Robert Shiller's *Animal Spirits: How Human Psychology Drives the Economy and Why it Matters for Global Capitalism* was published. This book provided an oven-ready explanation of the ensuing economic crisis. Akerlof and Shiller were two of an emerging band of economists who had taken a keen interest in the role of psychology and behavioural instincts within the operation of the economy. This group of scholars helped to form a novel approach to economic analysis that is commonly referred to as behavioural economics (Jones et al., 2013; Whitehead et al., 2017). While appearing to find its epistemological time in the financial upheavals of 2008 and 2009, behavioural economics had been an emerging area of inquiry for much of the second half of the 20th century.[2]

Despite its long antecedents, for much of its history behavioural economics was a fringe academic field that was shunned by the all-powerful principles of neoclassical economics and its associated neoliberal acolytes. Neoliberals did, of course, appreciate that humans did not always act in rational ways. But, according to Thaler, economists saw the irrational psychologies of human agents as 'true but unimportant': '[Neoclassical] [e]conomists were fine with the idea that their models were imprecise and that the predictions of those models would contain error. In the statistical models used by economists, this is handled by adding what is called an "error" term to the equation' (2015: 439). According to Akerlof and Shiller, the Credit Crunch and Great Recession represented the irrational remainder of the neoliberal economic equation. It embodied what happens when thousands upon thousands of microeconomic irrationalities came together. And it was not unimportant!

Interestingly, in the context of our discussion of smart technology, the neoliberal model of human behaviour had assumed forms of rational action that are more typical of a robot. What the behavioural economists argued was that we needed to replace *homo economicus* with a more realistic, flawed, and irrationally oriented account of human behaviour. At the onset of the Great Recession, Akerlof and Shiller thus asserted that '[w]e will never really understand important economic events unless we confront the fact that their causes are largely mental in nature' (2009: 230). For Akerlof and Shiller, these mental factors included confidence (and over-confidence), fear, and desire. They collectively referred to these factors as our animal spirits.

Drawing on decades of research into our animal spirits, behavioural economists offered a powerful diagnostic and cure for the problems of 2008. Diagnostically they argued that the economic meltdown of 2008 was not a random aberration within an otherwise stable socio-economic system. For behavioural economists, the Credit Crunch was the outcome of a series of predictable behavioural flaws which they had been assiduously diagnosing

---

[2]   The principles of what would become behavioural economics actually began in the 1940s and '50s with Herbert Simon's pioneering work on bounded rationality. It evolved through the work of Nobel Laureates Daniel Kahneman and Amos Tversky on cognitive bias in the 1970s, and was later advanced by prominent advocates such as Richard Thaler and Cass Sunstein (Sunstein would become the head of the Office for Information and Regulatory Affairs in the Obama Administration, and play a central role in applying the principles of behavioural economics to government policy; Whitehead et al., 2017).

for decades. Behavioural economists refer to these flaws as biases or heuristics (Kahneman, 2012). While they often lead to poor decision-making and harmful behaviours, biases and heuristics are an inevitable and necessary part of life. They reflect what Daniel Kahneman refers to as System 1 behaviours— or those behaviours that are made in quick and intuitive ways. A lack of time and the limited availability of information mean that we have to make hundreds of snap decisions every day. While some of these decisions are perfectly adequate, many do have deleterious consequences. The figure of *homo economicus* is, on the other hand, predicted on an assumption of System 2 behaviours: behaviours that are based upon reflection, calculation, and timely contemplation. The key contribution of behavioural economists has been to assert that System 1 is the dominant behavioural operating system for humans, and to uncover the consistency in the patterns of System 1 decision-making.

Common cognitive heuristics are present bias (or hyperbolic discounting), which is characterised by a tendency to prioritise the present over the future; social influence or herd bias, within which human behaviour is shaped by the actions of others; and the status quo bias, the human tendency to continue on an existing behavioural path as opposed to engaging in change. There are, of course, important evolutionary reasons why these behavioural biases exist: staying alive in hostile environments would appear to require some degree of prioritising immediate needs, staying within the herd, and avoiding risky changes in patterns of living. But in the modern world, behavioural economists argue that such biases were at the root of a range of intractable problems, including macrofinancial crises, climate change, personal debt, obesity, and political disengagement. It is clear, to be truthful, that some neoclassical economists were aware of the significance of more irrational forms of behaviour. In his influential analysis of the history of neoliberal thought, Michel Foucault (2008) observed that the exclusion of psychological insight from neoclassical economics was not because it was seen necessarily as insignificant. Rather, Foucault claimed that it was the unpredictable and inscrutable nature of behavioural psychology that made it incompatible with prevailing economic models of thought. By 2009, however, behavioural economists were confidently proclaiming both the importance and predictability of human irrationality.

If behavioural economics offered a diagnostic for the ensuing crises of humanity, the proposed cure would come from an engagement between behavioural economics and theories of cognitive design. This interdisciplinary interaction would be codified in what became the most influential behavioural economics book of all, *Nudge: Improving Decisions About Health, Wealth and Happiness*. Also published in the aftermath of the Credit Crunch in 2009, Thaler and Sunstein's *Nudge* suggested how policymakers (or choice architects) could creatively apply the insights of behavioural economics to a series of real-world problems. At the heart of the idea of nudging was the mobilisation of a form of soft power that relied on the reshaping of the contexts and so-called choice architectures of everyday life (Jones et al., 2011; Whitehead et al., 2011). Want to get people to reduce their levels of energy consumption? Tell them how their average energy use compares with their neighbours. Want to get more people to join organ-donor registers? Make it a prompted choice when the same people renew their driving licence. Want to get people to choose the best personal saving plan? Make it a default option. And, most famously, want to get people to eat healthier food? Then make sure that healthy food is at eye level in canteens. These subtle changes in our collective choice architectures don't force people to behave in a certain way (if you don't want to join an organ-donor register, just check the box 'no' when renewing your driving licence; if you don't want to choose the default saving options, then select another one; if you don't fancy that banana, then just reach over it to get the biscuit). Nudges were not designed to compel behaviour, but simply to make it easier for people to behave in ways that are most likely to facilitate improvements in personal wellbeing (Thaler and Sunstein, 2009).

The practices of nudging represent a creative fusion between behavioural economics and cognitive design. Behavioural economics identifies behavioural flaws, while cognitive design suggests ways in which we can design aspects of everyday life to exploit those flaws while also remedying them (for example, prompted choice requirements in organ-donor registers seek to address the status quo bias— simply not getting around to signing up—while also exploiting it: once on the organ-donor register, we are likely to stay there). Nudges are now used in a broad range of commercial and government spheres to try to shape human behaviour in a series of ways.

There are two important aspects of nudges that are particularly important to emphasise. First, nudges have gained widespread political support as a tactic because they are able to claim that they preserve freedom while still modifying behaviour. This suggests the possibility of a form of rebooted neoliberalism, whereby economic freedom is preserved while the behavioural shortcomings of the *homo economicus* model of human conduct are addressed (to put things another way, our collective problems are not caused by the free market, but the inability of humans to conduct themselves as effective free-market actors!). Second is the way in which nudges provide both behavioural explanations and solutions to socio-economic problems. Understanding socio-economic problems as behavioural tends to result in seeing structural issues as the aggregated outcome of individual actions. Understanding human conduct as primarily behavioural results in the use of behavioural solutions to socio-economic issues. These solutions do not focus on education or even persuasion (which target System 2 thinking), but rather the environmental factors (both social and physical) that condition the all-important realm of System 1 action. In the following section, we consider how the behaviouralist tendencies of nudges have resulted in an easy alignment with the behavioural project that is synonymous with smart technology.

Smart Tech Meets Behavioural Science—The Age of the Hypernudge

Although nudging has proved an increasingly popular approach to behaviour change and modification, it is marked by some clear limitations. Modifications to physical environments and choice architecture (such as streets and canteens) can present logistical and financial barriers to implementation. The most popular nudges have tended to be those that focus on the communication and framing of choice and prompts to action. Nudges have, for example, become popular in tax payment systems, where letters are used to convey social norms around paying your tax on time. The focus on low-cost, communication-based, and option-editing nudges has made them popular within austerity-oriented governments but has also limited their utility. Nudges have tended to be used to target one-off simple behaviours, such as joining an organ-donor register or selecting a particular mobile phone tariff. But nudges have been less able to shape more complex and continuous behaviours. There is also a tendency for the repetition of communication-based nudges to lose their impact over time (being told once that you consume more energy than your neighbourhood average may result in a re-think of your behaviour; being told the same message five times will probably mean you just ignore it). Beyond initial trials of nudging (randomised control trials are a methodology of choice), it is also very difficult to assess the impact of nudges at large scales and over long time periods. In these contexts, the early application of behavioural economic principles to human behaviour change have been characterised by what we would term a system of *nudge and go*: one-time nudges that target simple behaviours over a short time period.

Interestingly, while early strategies for applying nudges limited their impact on human behaviour, they also helped to allay ethical concerns that many had raised about the act of nudging (Jones et al., 2013; Whitehead et al., 2017; Furedi, 2011). By targeting more instinctive forms of human action, many had expressed concerns that nudges were forms of manipulation, even in their most benevolent manifestations. Furthermore, it was claimed that by deploying subtle forms of psychological power, nudges tended to subvert democratic norms and generate a new cadre of unaccountable choice architects (or *psychocrats*; see Jones et al., 2013; Legget, 2014). But if nudges are limited in scope and intensity, it appeared that these concerns could be easily allayed. The rise of smart technology has, however, opened up new opportunities for the governmental and corporate mobilisation of behavioural science. In doing so, it has also reenergised the debate about the ethical impacts of nudging on the human condition.

There are a series of very practical reasons why smart tech offers new opportunities for the mobilisation of nudges. First, the ubiquitous presence of smart technology in our everyday lives provides much more scope for continuous nudging at large scales. Second, it is much easier for choice architects to reshape the digital environments associated with online behaviours than to have to refashion physical space. Third, smart tech offers choice architects unprecedented opportunities to learn about the impact of nudges on behaviour. Smart technology is essentially a low-cost system for the delivery and monitoring of behavioural experiments. Fourth, smart technology enables far greater personalisation in nudges, as behavioural experiments reveal what works best not only at a population scale, but also for particular users (this is the basis of the smart-tech predictive system we outlined in the previous chapter).

The connections between nudges and smart tech run much deeper than these practical concerns. There are deeper normative connections between smart tech and the behavioural economics movement. These normative connections concern fundamental understandings of the human condition. While smart technology is heralded as offering the potential to greatly improve the human condition, its deployment has in part been driven by a belief in the inability of humans to live effectively in the big data age. There is simply too much to know, and too much to do. By drawing attention to the dominant role of System 1-type behaviours, behavioural economics depicts a similar vision of an enfeebled human state. Smart-tech advocates often see human limitations in technical terms (i.e. smart tech produces too much data for humans to be able to deal with effectively—thus only algorithmics will suffice). Behavioural economists are, however, more likely to see humanity's limits in evolutionary terms (our System 1 selves were central to our early survival, but now unhelpfully skew behaviour). There is, of course, a clear scope for alignment between these technological and evolutionary accounts of human limitations. In a 2009 article for the *American Psychologist* journal, Daniel Kahneman explains why behavioural economists value smart technology: '[t]he idea of algorithms that outdo human judges is a source of pride and joy for members of the heuristic bias tribe [namely behavioural economists]' (Kahneman and Klein, 2009: 523). While Kahneman acknowledges that algorithms do not, at the moment, outperform humans in all decision-making tasks, he suggests that they have great potential to help humans overcome their own cognitive failings. Thus, according to Kahneman, behavioural economists, 'are predisposed to recommend the replacement of informal judgment by algorithms whenever possible' (Kahneman and Klein, 2009: 518).

Algorithms have two properties that are of particular interest to behavioural economists. First, they have the ability to process large amounts of data (data that is often unavailable to humans, but even if it was, we could not process it quickly enough) in order to recommend optimal behavioural paths (perhaps the best route to take on a journey, or the ideal insurance package). Second, algorithms can process data at such rapid rates that they can be directly helpful when guiding the necessarily rapid decisions associated with System 1 processes. So, algorithms address the two main decision-making concerns of the behavioural economist: the limits of our behavioural environment and the cognitive-processing capacities placed on our decisions.

In the context of the practical and normative synergies between nudges and smart tech, it was only to be expected that the smart-tech society would witness the rise of new digital systems of nudging (see Weinmann et al., 2016; Yeung, 2016; Lanzing, 2019). These new forms of nudging are sometimes referred to as hypernudges or micronudges. While appearing to suggest two very different regimes of behavioural manipulation, the notions of hyper- and micronudging effectively convey the complex dynamics of the fusion of behavioural economics and smart tech. It was legal scholar Karen Yeung who first coined the term hypernudging. Yeung argues that,

> By configuring and thereby personalising the user's informational choice context, typically through algorithmic analysis of data streams from multiple sources claiming to offer predictive insights concerning the habits, preferences and interests of targeted individuals (such as those used by online consumer product recommendation engines), these nudges channel user choices in directions preferred by the choice architect through processes that are subtle, unobtrusive, yet extraordinarily powerful. (2016: 119)

In these terms, we can see that digital nudges are *hyper* to the extent that they are informed by unprecedented amounts of digital data about target subjects, and have the capacity to simultaneously reach and influence large numbers of people. Related nudges can be considered *micro* to the extent that they can be highly personalised and are able to permeate many aspects of our private lives. In her suggestion that hypernudges can be 'subtle, unobtrusive, yet extraordinarily powerful', Yeung (2016) draws critical attention to the ways in which digital nudges can overcome the constraints of analogue nudges yet still maintain their power over our collective unconsciousness.

We will reflect on Yeung's critiques of hypernudging later in the chapter. But at this point it is useful to introduce a more concrete example of a hypernudge in action. A much-discussed example of hypernudging is Facebook's Voter Megaphone project. The Voter Megaphone project is part of a series of initiatives that Facebook has trialled which seeks to target voter behaviour. Ostensibly, this initiative is not (necessarily) a partisan attempt to influence the way in which people vote, but an attempt to try and encourage more people to simply vote. In this context, the target behaviour is the casting of ballots, either digitally or at a polling station. The Voter Megaphone initiative was first deployed by Facebook in 2008 and has been used in a series of national elections around the world. While the Voter Megaphone initiative often takes slightly different forms in different elections, it is generally characterised by users being able to display an 'I've Voted' icon and to see who among their friends has cast their ballot. There are various aspects of this initiative that demonstrate the characteristics of hypernudging. First of all, it seeks to mobilise social influence (and our bias for social conformity) as a way of promoting voting action. Second, the Voter Megaphone project is able to mobilise social influence at very large scales (in the 2010 US congressional elections Facebook claimed it ran a trial of the Voter Megaphone with 61 million people; see Bond et al., 2012). Third, it is able to personalise influence (or micronudge) through its use of our own recognisable friends as social influencers. Fourth, it exhibits the experimental ethos of behavioural economists (in the 2010 trial, control groups were used to test the effectiveness of the initiative; Jones and Whitehead, 2018). The experimental insights gained from previous trials were then fed into subsequent iterations of the initiative.

There are other specific aspects of the Voter Megaphone hypernudge that are worth pointing out. Through its use of the 'I've Voted' emblem and our friend's faces, it used a nudging strategy that was fairly transparent. The Voter Megaphone project could be said to be conveying salient information about voter behaviour which could activate System 2 reflection on our own voting patterns (indeed, as part of the initiative, information was shared on where and how people could vote). However, through the use of social influence and the fact that the initiative did not seek informed consent from users to be part of the trial, it could be interpreted as an act of mass manipulation of System 1 behaviours.[3] As we will see in the next chapter, there is no reason why smart technology cannot be used to more effectively convey socially relevant information to make users better informed and more able to mobilise their agency. The question that we explore in the following sections of this chapter, however, focuses on the impact of smart tech's behavioural interventions and their likely implications for human agency.

---

[3]   Unsurprisingly, Facebook's Voter Megaphone initiative proved to be controversial. Despite its civic mission, there were concerns that through the selective use of control groups, it could be used to unfairly influence an election result. Facebook no longer uses this particular behavioural tool (Bolluyt, 2014). But what is not contested is the power that this particular hypernudge was able to wield. In a research letter published in the journal *Nature*, the team who ran the 2010 Voter Megaphone trial revealed that it increased voter turnout by approximately 340 000 people (Bolluyt, 2014: 296).

The Case for Hypernudging: Guiding Behaviours While Enhancing Freedom

So, the scientific rationale and technological capacity to hypernudge clearly exists. The question then becomes what are the justifications for applying hypernudges to shape human behaviour? The justifications for hypernudging are inevitably similar to ones that were used to justify earlier forms of analogue nudges (see Thaler and Sunstein, 2009). At the centre of these arguments is the question of the nature of freedom and its relationship to human behaviour. Freedom can be interpreted as a reflection of our ability to behave in ways that serve our interests (so long as the pursuits of those interests do not prevent others from pursuing their own interests, or cause harm to others). Freedom may also involve an ability to pursue our behavioural desires. Of course, behaving in ways that serve our best *interests* and our *desires* are often not the same thing. Overconsuming food may reflect an immediate desire, but it is clearly not in our best long-term interests. The capacity to effectively pursue our best interests and desires reflects what philosophers term positive freedom. Under neoliberal socio-economic systems, however, negative freedom has tended to be the predominant goal. Negative forms of freedom emphasise the importance of being left alone to determine our own behaviour—here, value is not placed on the capacity to behave in certain ways, but on preventing behavioural influence in our everyday lives.

Nudges and hypernudges are justified on the basis of two problems with negative freedom. First, it is claimed that negative freedom overemphasises the human behavioural capacity to achieve goals and desires in the absence of influence (indeed, it is argued that the pursuit of negative freedom merely exposes people to unregulated influences, particularly those emanating from the commercial sphere). Second, the pursuit of negative freedom makes it very difficult to tackle collective-action problems (such as climate change) effectively. Nudges and hypernudges are, however, also justified on the basis of limitations in the ways in which more positive approaches towards behavioural freedom have been pursued. Conventional support for the enhancement of positive freedom has focused on the personal capacity-building strategies of education and welfare. Behavioural economists and advocates of hypernudging argue that such approaches are not enough to unleash fully, positive behavioural freedom. For nudgers and hypernudgers, it is only when we take account of the cognitive limits that prevent us being able to realise our behavioural needs and desires that positive freedom will be achieved. It is thus claimed that hypernudges provide a system in and through which our capacity to act in our individual and collective interests is significantly enhanced (Sunstein, 2019). As we observed in relation to Facebook's Voter Megaphone initiative, this enhanced behavioural capacity is essentially pursued by utilising the very biases that often inhibit effective action.

The question of whether nudges and hypernudges are being applied ethically is determined by a series of considerations. First is the determination that a hypernudge is targeted at a beneficial behaviour around which there is a high degree of social consensus. Hypernudging people to vote would in this instance be a very different proposition than nudging people to invest their savings in the stock market. Of course, this ethical test is not easily resolved. While many surveys demonstrate that people want to join organ-donor registers, nudging people onto them has generated significant resistance from religious groups who oppose organ donation. It is also clear that not all legitimate interventions in human behaviour have to be granted consent by the public (paying taxes may struggle on this front; House of Lords, 2011). An important aspect of an ethically designed nudge is also that it should be relatively easy to resist (see Thaler and Sunstein, 2009). The human ability to override a hypernudge is a design element that seeks to preserve human behavioural autonomy (resisting Facebook's Voter Megaphone initiative is relatively simple: don't vote). In this context, nudges are often seen as expressions of libertarian paternalism. Libertarian paternalism is an approach to human behaviour change that seeks to enhance human welfare (the paternalist aspect), while not undermining personal freedom (the libertarian dimension).

Although libertarian paternalism has been described by some as an oxymoron, it does demonstrate how hypernudges are trying to enhance positive aspects of freedom while not undermining negative freedom. The extent to which hypernudges are able to balance libertarian and paternalist goals is an open question. What is harder to dispute is that the nudging agenda reflects a significant intellectual challenge to established ways of understanding behavioural freedom (see Sunstein, 2016, 2019). It ultimately seeks to address the behavioural challenges that exist to achieving positive forms of freedom by reframing choice. Choice (both in terms of framing and preservation) is the ethical hinge upon which hypernudging hangs. As we discuss in greater detail in the following section, the seemingly banal behavioural reinterpretation of freedom as choice carries with it significant implications for how we understand the human subject more generally.

Another central aspect of the ethical application of hypernudges is their level of transparency (House of Lords, 2011). Transparency is important to the ethics of nudging in general terms because it affords a degree of human capacity to choose to resist behavioural influence. Given that many hypernudges are being conducted by private sector companies, transparency becomes particularly significant. While we expect that government nudges should be subject to conventional forms of political scrutiny, corporate hypernudges are likely to be far less regulated. Thus, the ability of individuals to clearly see hypernudges would appear central to protecting us against unwanted forms of behavioural manipulation. Transparency in hypernudges can be achieved in two main ways: (1) directly informing people of the nudge; and (2) making it possible for people to easily discern the presence of a nudge (House of Lords, 2011). Given that telling people that they are being nudged tends to diminish its behavioural power, it appears that the key to transparent hypernudging is reasonable discernment. Determining thresholds for reasonable discernment is difficult when applying analogue nudges. In the context of the reach, scope, and variations of behavioural prompts associated with hypernudging, maintaining this ethical standard could be nearly impossible.

Having established the ethical case and capacities for hypernudging, it is important to reflect on their broader social, economic, and political benefits. In a piece written for the *Behavioral Scientist* in 2017, the then Head of Behavioural Design at Capital One, Chris Risdon, outlined the broader significance of hypernudging. In the context of trying to promote saving behaviours within his bank, Risdon (2017) saw real scope for the smart-tech-informed application of behavioural science. According to Risdon (2017), the combination of data science and behavioural science embodied in the hypernudge has three main benefits:

1. It enables strategists to mathematically discern optimal saving behaviours for specific users (in terms of account type, saving amount, and timing);
2. It enables the calculation of optimal saving rates according to the particular needs of an individual (on a particular income, with specified outgoings, and particular saving goals); and
3. It is able to use these insights to scale related behaviour changes to millions of users.

It is the particular combination of mass-scale behavioural modification with individual specificity (or persuasion profiling, as Risdon describes it) which Risdon argues marks hypernudging out. According to Risdon,

> The implication is that if people invite positive influence, we need to scale our ability to influence them, not just through marketing and acquisition, but the entire product or service lifestyle ... It's a two-part challenge. Augment the person's rational self by providing a decision engine that, using machine learning, finds insightful patterns in their life ... Then, control a person's irrational self: identify and target their unique combination of biases and heuristics they use to make decisions, and protect them from those biases. (2017: 3)

For Risdon, then, the fusion of data and behavioural science embodied in the hypernudge enables a synchronised enhancement of humanity's System 2 rationality and a holding at bay of our worst behavioural System 1 instincts.

In terms of behavioural modification, Risdon goes as far as to argue that hypernudging reflects a new technological age:

> We are moving from an age in which products and services connect us to and better manage our things (music, money, email, friends) to an age in which products and services are explicitly designed to help us achieve behavior-based goals. In other words, we are moving from the utility age to the augmentation age. (2017: 2–3)

This shift from utility to augmentation is a shift from smart tech making our lives easier to smart tech shaping our life paths in the direction we (may) wish to travel. In terms of banking and personal finance, Risdon suggests that hypernudging can initially help people to manage their money more effectively. But in the longer term these nudges will also facilitate the realisation of other behaviours, including earlier retirement or the buying of a house. He even speculates on the behavioural augmentation prospects of Google Calendar's *Google Goals*, and its potential to manage our time and enable us to realise new ways of achieving our desired work–life balance.

Within his analysis of digital nudging, Risdon offers some speculative reflections about the potential to apply hypernudging in real-world settings. Reflecting on Amazon's prototype checkout-free stores, he suggests that the flexible digital spaces of our online existence could soon be reflected in the spaces of everyday life. This would, of course, enable hypernudging to exploit a new set of choice architectures and influence ever more deeply our behavioural destinies. Ultimately, and perhaps with a heavy dose of hyperbole, Risdon sees hypernudges as opening up a new era of behavioural modification:

> Just like unlocking the human genome helped identify genetic traits that allow for personalized medical advice, we can think of machine learning as the next step in unlocking a 'behavior genome'. By factoring in personality traits, situational features, and timing, we can better persuade people who want to be persuaded. (2017: 6)

It appears that there are two arguments in favour of the application of hypernudges. The first rests on a realisation that human decision-making is flawed and in need of consistent support to enable us to realise meaningful forms of freedom. The second argument rests on the potential for smart technology to meaningfully augment human decision-making and behaviour in unprecedented ways. We now turn our attention to a series of counter arguments that focus on the real and potential problems of smart tech's emerging behavioural interventions.

CRITICAL PERSPECTIVES ON SMART TECH'S BEHAVIOURAL PROJECT

Hypernudging and the Schizophrenic Subject

In addition to identifying and outlining the concept of the hypernudge, Karen Yeung offers us one of the most comprehensive critical analyses of its implications. Yeung (2016) describes her assessment as a liberal rights-based critique. At the heart of Yeung's analysis is a concern with the implications of hypernudging's digital decision guidance processes for human flourishing and democracy. Yeung builds on existing critiques of static nudges to consider the novel challenges that more dynamic digital nudging regimes present to the human condition. In this context, Yeung seeks to bring attention to the implications of hypernudging for 'liberal democratic principles and values rooted in the respect for individual autonomy' (2016: 123). Ultimately, Yeung demonstrates that hypernudging depends on a peculiar maintenance of two contradictory visions of human autonomy, which suggest the presence of a decidedly schizophrenic subject.

Yeung's assessment rests on two key differences between analogue and digital nudging. The first are the power asymmetries that exist between 'global digital service providers, particularly Google and Facebook, and individual service users' (Yeung, 2016: 123). According to Yeung, while knowledge asymmetries exist within static nudging regimes, these are dwarfed by the surveillance capacities of Big Tech. This imbalance in knowledge is significant not just because it reflects an uneven distribution of power, but because this power is being accumulated by politically unaccountable private corporations. The second difference relates to the scales of influence associated with hypernudging. While Risdon (and other behavioural designers) see the scalability of hypernudges as one of their primary advantages, for Yeung, hypernudgers' ability to 'directly affect millions of users simultaneously' (2016: 123) makes them a real and present threat to liberal democracy. For Yeung, the power asymmetries and scalability of hypernudges means that critique should no longer be focused on questions of efficacy (as it was in relation to analogue nudges)—the effectiveness of hypernudges is demonstrable. Instead, Yeung (2016) argues that critical analysis should instead focus on the extent to which hypernudges embody forms of covert manipulation, and what this means for human behaviour and democratic freedom.

There are three reasons why hypernudges can be considered acts of covert manipulation, as opposed to benevolent acts of behavioural augmentation (Yeung, 2016: 123–4). The first is the evident use of hypernudges for what may be considered illegitimate purposes (123). This critique takes us back to our previous discussion of the ethics of behavioural nudges being, in part, determined by the extent to which they are associated with the promotion of beneficial behaviours, around which there is a high degree of social consensus. Yeung argues that as hypernudging is largely driven by corporate rather than governmental actors, the social value of the behavioural outcome of nudging is likely to be much more contested. Drawing on the example of Facebook's controversial Emotional Contagion Trial (within which the feeds of approximately 700 000 users were altered to analyse the impacts that this had on people's emotional states), Yeung claims that there is a danger that hypernudges simply become 'mass experiments in emotional manipulation' (2016: 123). The illegitimacy of Facebook's Emotional Contagion Trial (at least in the context of liberal democratic norms) stems from two primary features: (1) the fact that it did not seek the active consent of users who were enrolled in the trial; and (2) the fact that it was not based on a clear benefit to users (beyond the rather self-serving claims of service improvement offered up by Facebook; Yeung, 2016: 123).

The second reason that hypernudges can be thought of as acts of pernicious manipulation is because of their deceptive qualities (Yeung, 2016). To the extent that all nudges target human cognitive limitations they can all, to some extent, be considered acts of deception. Accusations of deception are generally dispensed with by advocates of analogue nudging on the basis that the behavioural target is one that is established by accountable political systems and serves clear public benefits. Hypernudges are more problematically deceptive to the extent that they serve much narrower corporate purposes. In this context, the distinction between nudges and hypernudges is as much a product of who the choice architect is (the public or private sectors) as opposed to the novel technical capacities of digital nudging. Drawing on the work of Pasquale and Bracha, Yeung (2016) demonstrates the deceptive qualities of hypernudging through the example of the way in which search engines rank websites (Pasquale and Bracha, 2015; Pasquale, 2015). Search engines such as Google use website ranking to steer and nudge us towards particular websites. While these rankings are, of course, designed to make our navigation of the internet easier, they are also influenced by corporate interests who are keen to promote their own particular products. According to Yeung, then (and following Pasquale and Bracha, 2015),

> search engines filter and rank websites based on criteria that will inevitably be structurally biased (designed to satisfy users and maintain a competitive edge over rivals), thus generating systematically skewed results aimed at promoting the underlying interests of the gatekeeper, thus distorting the capacity of individuals to make informed, meaningful choices and undermining individual autonomy. (2016: 124)

Big Tech's hypernudges are thus deceptive to the extent that they seek to shape human behaviours in ways that are not necessarily in people's best interests, and in ways that users are not able to discern. In this context hypernudges are not about enabling people to control their inherent biases in navigating data, but about introducing corporate biases into smart-technology functions. The act of deception is thus realised in the presentation of search results as being optimal for the users, when in fact they are actually optimal for corporate choice architects. It is about the deliberate concealing of information that could have been valuable in making informed decisions.

The third reason that Yeung argues hypernudges embody problematic forms of behavioural manipulation is because of their associated lack of transparency. As previously mentioned, transparency (or an ability to see and question the operation of nudges) is central to the legitimate use of nudge techniques in liberal democracy. Without some degree of political transparency there is no guarantee that nudges will not simply operate on the collective unconscious without people's awareness or consent. Concerns have been raised about the transparency of analogue nudges, with some arguing that they rely on citizens having to place too much trust on the benevolent intents of avuncular choice architects. But, according to Yeung, hypernudges are associated with new problems of transparency. Yeung observes how 'the critical mechanisms of influence utilized by hypernudging are embedded into the design of complex, machine-learning algorithms, which are highly opaque' (2016: 124). Not only are the biases and the deeper purposes of hypernudges obscured by their embedding in inaccessible computer code, these very codes and algorithms are also protected as proprietorial assets. Amoore (2020) astutely reminds us that the problems of the smart-tech society cannot be solved by simply seeing and fixing algorithmic code. Nevertheless, the inability to effectively see the operative principles of hypernudges makes them antithetical to many of the established norms of liberal democratic society.

Having outlined the novel dilemmas that hypernudging presents to human autonomy and democratic norms, Yeung considers how smart-technological acts of behavioural manipulation have been justified. It is here that we encounter the most significant implications of Yeung's work. According to Yeung, the spread of hypernudging (and indeed the broader forms of surveillance and data gathering associated with smart technology) have been justified on the basis of regulatory systems that emphasise consent and privacy self-management. As opposed to more draconian regulatory measures, in liberal democracies the activities of smart-tech companies are generally governed by systems of notice and consent (Yeung, 2016: 125). Notice and consent systems rely upon the assumption that the users of smart tech can give meaningful consent for the collection and use of their online behavioural data following notification. The act of notification of data use is the, admittedly limited, moment of hypernudge transparency. Notice and consent systems are seen to be compatible with liberal democratic norms to the extent that they are associated with the provision of information upon which users can make and enact autonomous behavioural decisions. Yeung outlines the various limitations that are associated with notice and consent systems. First is the evidence that people do not read or meaningfully engage in smart-tech consent notices. In this context, Yeung reflects upon the research of McDonald and Cranor (2008), who indicate that if users read all of the digital consent notifications they received, it would require on average 244 hours of dedicated reading a year! (Yeung, 2016: 125). There is of course a paradox here: smart tech is increasingly being used to help people deal with their cognitive limitations, and yet its legitimate use relies on active engagement with terms and conditions that people simply do not have the time (and often specialist knowledge) to effectively engage with. Could this be another case of Big Tech using our bounded rationality against us? Second, Yeung highlights that giving meaningful consent to be hypernudged is made nearly impossible because of the uncertainties that surround the long-term utility of our online behavioural data (2016: 125). Smart tech's use of our behavioural data is most powerful when it is based on the long-term accumulation of data across various platforms. This process makes it impossible for consent to be given to hypernudgers in anything but the most generic of ways. In other words, even if we had the cognitive bandwidth and time to effectively process digital consent notices, it is highly unlikely that we could consent to specific hypernudges, as the form these could take would most likely not be known in advance.

*

Ultimately what Yeung's analysis demonstrates is an exploitation of privacy behaviours by Big Tech. The notice and consent system does not enable users to take meaningful control over their behavioural data, while hypernudges can be easily deployed to militate against people enhancing their privacy practices (2016: 125). Ultimately what appears to be emerging in the smart-tech society is a strangely schizophrenic understanding of human behaviour. On the one hand is the irrational limited subject which smart technology seeks to support (and exploit) through systems of hypernudging. On the other hand, there is the enduring figure of *homo economicus* who is invoked by regulatory authorities to support a belief in the provision of meaningful consent to highly complex and behaviourally manipulatively smart-tech systems. At the very least it would appear that new understandings of the constraints that exist to being able to give consent to behavioural modification should be developed within the regulatory regimes associated with smart tech. Yeung also suggested that more attention be given to the long-term impacts that smart-tech systems and hypernudge initiatives are having on the behavioural patterns and capacities of users. If the message of behavioural economists is true and we are, behaviourally speaking, the products of our contexts, it is important to ask what types of subjects smart-tech environments produce. For Yeung, the stakes are high and pertain directly to the future of democratic self-government:

> Before succumbing to the allures of the convenience and efficiency that big data claims to offer, we must be attentive to its regulatory power, operating as a particularly potent, pervasive yet 'soft' form of control, modulating our informational environment according to logics that are ultimately outside our control and which erode our capacity for democratic self-government. (2016: 132)

Surveillance Capitalism, Behavioural Actuation, and Algorithms as Choice Architects

In her previously discussed analysis of surveillance capitalism, Zuboff (2019) also developed a critical analysis of the behavioural dynamics of smart technology. While this critique overlaps with some of the issues raised by Yeung, Zuboff developed novel concerns pertaining to the impacts of behavioural modification. According to Zuboff, the behavioural impacts of smart tech reflect a new phase within the predictive imperative of surveillance capitalism (see Chapter 3 of this volume). Zuboff describes this phase as the 'economies of action'—the profits that can be made from converting predictions of human desires into action (or what specialists term actuation). Although converting predictive insights about human need into related behaviours is an expected economic development, Zuboff raises concerns about what behavioural actuation means for human subjectivity. If, as we established at the beginning of this chapter, the self-authorship of our behaviours is central to human self-identity, what does mass manipulation mean for humanity?

In her critical analysis of behavioural actuation, Zuboff charts the connections between behavioural economic thinking and smart technology. Zuboff states,

> Surveillance capitalists adapted many of the highly contestable assumptions of behavioural economics as one cover story with which to legitimate their practical commitment to a unilateral commercial programme of behaviour modification. The twist here is that nudges are intended to encourage choices that accrue to the architect, not the individual. The result is data scientists trained on economies of action who regard it as perfectly normal to master the art and science of the digital nudge for the sake of their company's commercial interests. (2019: 295)

Zuboff identifies three main mechanisms through which the power of nudging is realised across smart-tech platforms. The first is the act of tuning—that is, the use of subliminal clues and psychological prompts to action. The second are acts of herding, or the 'remote orchestration of human behaviour by controlling context' (Zuboff, 2019: 295). In the context of herding, Zuboff describes how the spread of ubiquitous computing into our cars, homes, and workplaces means that smart tech now has the ability to more coercively shape human action. To demonstrate the emerging potential of herding techniques, Zuboff quotes from an interview she conducted with an IoT software developer:

> We can engineer the context around a particular behaviour and force change the way context aware data allow us to tie together your emotions, your cognitive functions, your vital signs etcetera. We can know if you shouldn't be driving, and we can just shut your car down. We can tell the fridge, 'Hey, lock up because he shouldn't be eating', or we tell the TV to shut off and make you get some sleep, or the chair to start shaking, because you shouldn't be sitting so long, or the faucet to turn on because you need to drink more water. (IoT Software developer, quoted in Zuboff, 2019: 295–6)

This dystopian vision clearly takes us beyond the realm of nudging behaviours into digital coercion. But compared with tuning, these acts of coercive paternalism do at least enable those who are subject to them to see the ways in which their behaviours are being influenced, and potentially resist them.

The third behavioural actuation tool identified by Zuboff is that of conditioning. The process of behavioural conditioning derives directly from the work of the prominent behavioural scientist B.F. Skinner. Through his work on laboratory rats and pigeons, Skinner claimed that all animal behaviour (including that of humans) was primarily the product of the forms of positive and negative reinforcement we get from our surrounding environment. Receiving good feedback (in the form of a reward, perhaps) makes that behaviour much more likely to be replicated in the future; receiving an unwanted response to an action means that that behaviour is likely to be suppressed. Of course, smart technology provides us with an unprecedented set of opportunities to provide real-time feedback and reinforcement for our behaviours. Indeed, many smart devices use rewards systems (such as digital trophies and celebratory graphics) to provide immediate feedback on our actions. Although the use of digital conditioning may have an innocent air, it is important to reflect on the underlying assumption upon which it is based.

The publication of Skinner's 1972 *Beyond Freedom and Dignity* was an attempt to explore the implications of his work for society. *Beyond Freedom and Dignity* embodied a direct behaviourist attack on liberal assumptions of subjectivity. In this controversial volume, Skinner targets *homo politicus* (in his words, the 'inner man', or 'autonomous man' [*sic*]) as the hollow subject of Western philosophy and democracy. According to Skinner, 'The function of the inner man [*sic*] is to provide an explanation which will not be explained in turn. Explanation stops with him. He is not a mediation point between past history and current behaviour, he is a centre from which behaviour emanates' (1972: 14, emphasis in original).

Furthermore, Skinner observes, 'He [the inner man] initiates, originates, creates, and in doing so he remains, as he was for the Greeks, divine. We say he is autonomous—and, so far as a science of behaviour is concerned, that means miraculous' (1972: 14). On these terms, acts of behavioural conditioning are not just sets of innocent nudges; they collectively embody a vision of a post-Enlightenment human condition, a recognition that humans are not endowed with a special inner purpose and agency, but are merely responsive creatures like all animals.

Significantly, Zuboff (2019) argues that behavioural actuation embodies a shift in the predictive dynamics of smart technology. Making specific reference to Facebook, Zuboff observes that, 'Facebook's surplus is aimed at solving one problem: where and when to intervene in the state of play that is your daily life in order to modify your behaviour and thus sharply increase the predictability of your actions now, soon, and later' (2019: 300). You will notice here the distinction between prediction and predictability. Predicting what you may need reflects smart technology's ability to analyse patterns in previous behaviours. 'Increasing the predictability' of human behaviour is an altogether different thing. Zuboff appears to be claiming that by deploying the varied psycho-technological tools of hypernudging, smart-tech platforms are trying to generate predictability in human responses to prompts, rather than identifying existing human needs. This is the difference between predicting when someone is going to like or need a product or service and actually being able to stimulate that need. This is behavioural conditioning par excellence. To demonstrate the forms of behavioural conditioning deployed within the smart-tech world, Zuboff draws on the example of Pokemon Go. This popular game has been experimenting with the most effective techniques that can be used to digitally drive people to visit specific sponsored sites as part of the game (such as McDonald's restaurants). What is, perhaps, most interesting about Pokemon Go, though, is not that it is an ongoing experiment in behavioural modification at scale (namely, the modification of routing and walking behaviours). As we would expect from smart technology, the Pokemon Go experiments in behavioural conditioning are automatically generated. Different prompts are used at different times to test their efficacy in an almost infinite loop of behavioural nudges and machine learning. In this context, it is an algorithm (or interconnected set of algorithms) that acts as the choice architect of hypernudges. It seems unlikely that algorithmic choice architects are likely to display the commitments to ethics and transparency that are central to the legitimate use of nudges in free and democratic societies.

Perhaps the most telling aspect of Zuboff's analysis of digital nudging is her reflection on its unconscious nature. Previously in this chapter we spoke about how nudges tend to work best in the dark spaces of the human unconscious, but that this does not mean that they cannot have transparent elements. In relation to the actions of surveillance capitalist firms, however, Zuboff argues that the operation of nudges in the unconscious realm has commercial significance. Zuboff states, '[S]urveillance capitalists' ability to evade our awareness is an essential condition for knowledge production' (2019: 328). For Zuboff, then, the principle that behavioural modification should evade human awareness is not just about maximising the efficacy of digital nudges; it is central to the unregulated expansions of surveillance capitalist enterprises. Zuboff also observes that, 'surveillance capitalists declare their right to modify others' behaviour for profit according to methods that bypass human awareness, individual decision-rights, and the entire process of self-regulatory processes that we summarise with terms such as autonomy and self-determination' (2019: 298).

If Zuboff is correct, then smart tech is not using behavioural conditioning because it enhances frayed forms of human autonomy and compensates for the unrealistic expectations of self-determination associated with *homo economicus*. Behavioural conditioning (and its particular manifestations that are favoured by surveillance capitalists) are used because they enable smart tech to deliberately avoid human regulation. Again, returning to Zuboff:

> The evasion is neither accidental nor incidental, but actually essential to the structure of the whole surveillance capitalist project. Individual awareness is the enemy of telestimulation because it is the necessary condition for the mobilisation of cognitive and existential resources. There is no autonomous judge without awareness. Agreement and disagreement, participation and withdrawal, resistance and collaboration: none of these self-regulating choices can exist without awareness. (2019: 307)

These may seem like heady claims, but they raise a troubling prospect. Many, including ourselves, have critiqued Zuboff's analysis because of its dependency on humans actually being the easily manipulable animals that Skinner and the smart-tech sector seem to assume (see Whitehead, 2020). But what if the issue was not so much whether humans lack the autonomous capacities and intentionality to resist smart tech's manipulative overtures? What if the behavioural project of smart tech—through the unprecedented scales of its behavioural trials and its ability to evade human awareness and resistance—is actually supporting the creation of the flawed human subjects described by behavioural economists? What if the smart-tech society is not so much helping us deal with our System 1 problems, but actually eroding our System 2 capacities for deliberation and autonomous action? As Zuboff states, '[e]very threat to human autonomy begins with an assault on awareness' (2019: 308).

CONCLUSION

In 1998 B.J. Fogg wrote an influential paper titled 'Persuasive Computers: Perspectives and Research Directions'. Fogg was working at Stanford University and for Sun Microsoft Systems. As a behavioural scientist Fogg was one of the first people to systematically assess the capacity of computer technology to consistently shape human behaviour. Fogg is of particular importance to the themes covered in this chapter because in 1997 he established the Stanford Persuasive Technology Lab, later renamed the Behaviour Design Lab. This lab has played a central role in ensuring that generations of those working on the development of smart technology are well versed not only in computer code but also in behavioural science.

From its inception, Fogg and the Behaviour Design Lab have emphasised the potential utility of persuasive technologies as well as the ethical dilemmas they are likely to generate. In his first published piece on persuasive technology, Fogg observed that

> [t]he study of computers as persuasive technologies raises important ethical questions. This is understandable and inescapable, given the fact that persuasion has long been an area for ethical debate, and that computer technology has raised recent ethical questions. As with any other means of persuasion, one could compromise values and ethics intentionally or unintentionally. Therefore, those who study persuasive technologies should have a sound understanding of the ethical implications of this field. (1998: 231)

In a *Wired* interview in 1997, Fogg draws particular attention to the importance of developing human awareness of the dangers of technological behavioural manipulations; 'Fogg says as the internet grows as a vehicle to change what people do, people will have to start paying attention in order to protect themselves' (Wired Staff, 2000).

In this chapter we have seen that as smart technology becomes more ubiquitous, the potential that Fogg saw in persuasive technology has grown immensely. This potential has been super-charged by new breakthroughs within the behavioural sciences and the emergence of nudge techniques. We have also seen how the ethical concerns identified by the Behaviour Design Lab have come to fruition. In an age of surveillance capitalism, it appears that Fogg's call for vigilance may be being actively undetermined by the smart-tech industry's bypassing of human awareness. Ultimately, this chapter has illustrated that our attention should not so much be on the impacts of smart tech on behaviour, but *how* smart tech seeks to shape behaviour. Smart technology can clearly be used to enhance positive freedom, augment human capacities, and help us build healthier and more fulfilling lives. At the same time, however, persuasive digital technology can also be used to undermine human freedom, manipulate human behaviour for corporate benefit, and ultimately diminish the human condition.

In this chapter we have primarily focused on the fusion of smart tech with hypernudge techniques because they deploy forms of influence that are most compatible with the norms of liberal societies. Digital behavioural influence is, however, being used in more coercive forms in many parts of the world. With the emergence of Bitcoin and block-chain-based central digital currencies, the potential for more coercive behavioural manipulations in liberal and authoritarian states will only grow. Reflecting on the connections between central digital currencies and COVID-19 policy, Chapman, for example, emphasised the behavioural potential of virtual currencies:

> Imagine if, after a future economic shock (such as another pandemic) the government issued support in the form of a digital-currency credit, instead of handing out benefit payments in cash that might be hoarded, gambled or spent on alcohol. This CBDC [Central Bank Digital Currency] could be programmed to expire in three months and only be spent on certain 'socially desirable' goods or services—one example of how CBDCs could be used to restrict liberty and influence behaviour through social engineering. (Chapman, 2021: 38)

Yet, despite such coercive potential, even such acts of digital-currency-based manipulation are perhaps less troubling than the unconscious conditioning described by Zuboff. Forcing people to spend money on certain things in certain time periods at least enables people to be aware of what is happening and to contest behavioural manipulation politically.

Our focus in this chapter has also meant that we have not considered perhaps the most significant behavioural goal of smart technology: enhanced and continued engagement with smart-technology devices and platforms. The behavioural target of technological engagement utilises nudges, but also deploys the addictive arts of marketing. Addictive behaviours are pursued through the promotion of controversial and divisive content. This also involves the use of dopamine-rich engagement environments that use variations and rewards to promote continuous engagement. Given that smart tech's ability to change our behaviours for good or ill relies on these forms of primary user engagements, the human capacity to regulate our interaction with smart tech is perhaps the primary behavioural issue of our technological age.

5.   The smart body—from cyborgs to the quantified self

INTRODUCTION

Our discussions in the previous two chapters concerning smart-tech predictions and behavioural influence have drawn attention to the ability of digital technology to learn more about humans as social beings. One of the most significant areas of recent smart-technology development has, however, been centred on the human body (Dow Schüll, 2016). At the core of this project is a desire to better understand humans as biological entities. Related developments reflect a kind of biological dimension of our *data selves* (Cheney-Lippold, 2017). For Cheney-Lippold, our data selves are algorithmic projections of our personalities, predilections, and anticipated behaviours. The formation of our bio-data selves, however, operates at a more embodied level. It also involves a shift away from more intuitive regimes of biological understandings of our bodies. Instead of aches and pains, tiredness, 'feeling under the weather', or just 'feeling fine', we have precise biometric diagnoses of the biophysical fluctuations of our bodies (Davidson, 2016). Instead of what Davidson calls a 'tech-free sensing self', we have a biometric entity that can be effortlessly compared with other bio-data selves and algorithmically diagnosed for wellbeing or disease (2016: 28).

It is something of an irony just how little we know about what is going on in our body. Despite being an intrinsic part of every aspect of our existence, the body always exceeds our ability to perceive it fully and accurately. When we refer to the smart body, we are referring to myriad ways in which our bodies are now being incorporated into the digital feedback loops associated with smart technology. The smart body is thus characterised by the ways in which certain aspects of outer-bodily appearance can be used to identify us, learn about us, and trigger feedback and behavioural prompts that are tailored for us. But the smart body is also a product of wearable technologies that can be used to measure various proxies for the internal health of our body and its organs. In this context, smartwatches can now measure heart-rate levels (and provide health warnings related to irregular heartbeat patterns), exercise routines, walking steadiness (and likelihood of falling), and sleep patterns.

Mobile phone apps now enable people to more accurately record the fluids, fibres, vitamins, and minerals that go into our bodies. There are now even 'internet-enabled fabrics' that can be used to sense changes at the surface of our skin (Zuboff, 2019: 246). The collection and analysis of this bodily data offers unprecedented opportunities to gain insights into personal and public health patterns.

As is the case with all manifestations of smart technology, the rise of the smart body is associated with certain costs and potential benefits. The smart body reflects a chance to know our biological selves in novel ways, and for parents and doctors to be better able to anticipate the health problems and needs of those in their care. At an individual level, the smart body and the biological feedback loops it can provide us create a new motivational context within which we can attempt to achieve and monitor improvements in personal health. But smart body technologies, and the biometrics they produce, are also being exploited for commercial gain and political expediency. In its most pernicious form, perhaps, Zuboff outlines the ways in which smart body technology is being deployed within China's Social Credit System as a way of enforcing a kind of 'digital totalitarianism' and 'total (human) awareness' (2019: 388–94).

While we share Zuboff's biometric concerns, in this chapter we engage in a fairly open exploration of the emergence of the smart body and its associated potential. Through a form of auto-technological ethnography, we draw attention to the lived experience of trying to create a smarter body. While we highlight the darker side of smart body technology, we take seriously the benefits it can bring and draw attention to how these benefits can be more reliably achieved. This chapter begins by considering the history and nature of smart body technologies. We then move on to consider existing interpretations of these technologies and their consequences for human health and social systems more generally. The main section of this chapter introduces an auto-technology ethnographic. This ethnography recounts the experiences of one of the authors of this volume (Whitehead) as he utilised various smart body technologies over a three-week period. This auto-ethnography offers insights into the positive and negative aspects of smart body technologies from the perspective of the everyday experience of a user.

THE HISTORY AND DEVELOPMENT OF SMART BODY TECHNOLOGIES

The current significance of smart body technologies can, in part, be perceived on the basis of the economic value of the sector. Smart body technologies can be broadly divided between wearable technologies that monitor individual bodies, and the broader sweep of biometric systems which can use various parts of the body for identification purposes. In 2020 the value of the global biometrics market was estimated to exceed £20 billion (Global Market Insights, 2021). Global Market Insights estimate that there will be a significant growth in the use of biometric systems over the coming decade. They claim that digital systems that record various biometrics (including face, iris, palm and fingerprints, and voice) will experience a market growth of approximately 13 per cent (Compound Annual Growth Rate) between 2021 and 2027 (Global Market Insights, 2021). Biometrics technologies will play an increasingly important role in facilitating access to various digital platforms and resources. One of the major reasons for the growth in the biometrics technology market is the desire of governments to move towards e-government systems (Global Market Insights, 2021). In this context, it is claimed that the ability to gather and monitor citizens' biometric data will play a crucial role in facilitating quick and secure access to government services (see Chapter 7, this volume). Singapore's government is, for example, deploying facial recognition technology as part of its national identity system (Global Market Insights, 2021). Although biometric technologies do appear to offer precise identification systems, they are routinely criticised as an intrusive form of digital surveillance that supports dehumanised e-government and e-commerce systems (see Chapter 7, this volume).

The extent to which biometric identification systems can be considered a smart technology is debatable. It is certainly possible that the routine identification of people through various forms of unique body identification could be utilised for digital learning and the provision of behavioural feedback or tailored services. But, while potentially making our lives easier, the use of biometrics merely for rapid identification should not in itself be considered a smart technology. The use of wearable technology tends to be more consistently aligned with smart learning and feedback, however. At $40.65 billion, the size of the wearable technologies global market was approximately twice the size of that associated with biometrics in 2020 (Grand View Research, 2021). As with the biometrics sector, wearable tech is expected to see significant market growth over the next decade (Grand View Research, 2021). The growth of wearable smart technology appears to be related to a series of trends including concern over chronic illnesses related to obesity and the rise in popularity of such devices among fitness enthusiasts (Grand View Research, 2021). It is claimed that the increasing health awareness generated by the COVID-19 pandemic is likely to lead to an increased use of such devices over the coming years.

In their early iterations, wearable technology and fitness trackers were not smart. They could record simple things, such as step counts, but offered little in the way of digital learning or bespoke feedback. With the emergence of market leaders, such as Fitbit, Garmin, and Xiaomi, the amount of data that such technologies can record and share has expanded significantly. The latest smart wearable technologies can automatically record floors walked (altitudinal change), heart rate, locational data, and quality of sleep. Many devices also enable users to manually record the food and liquids they consume throughout the day as well as wellbeing indicators (Grand View Research, 2021). Through Wi-Fi connections to mobile phones and uploads to websites, related bio data can be automatically analysed and compared to historical data to identify trends and performance patterns.

The sites of bodily connections between humans and smart tech are becoming increasingly diverse. Although wristwear tech still dominates the market in 2020 (with a 48.9 per cent market share), smart tech can now take neckwear, eyewear, headwear, and bodywear forms (wearable sweat sensors, for example, can now alert athletes to water and electrolyte losses; Grand View Research, 2021). Of course, not all smart wearables are focused on the monitoring of the human body. Smart eyewear and goggles, such as Google Glass and Microsoft HoloLens, can be used to orientate and guide people through space as well as provide easy access to relevant online tools and resources (Grand View Research, 2021). Indeed, smart goggles are increasingly being fused within the augmented and virtual-reality software emerging from the gaming sector to provide smart training devices for surgical practices (Grand View Research, 2021). We will discuss the application of wearable technology in the workplace in greater detail in Chapter 6. In this chapter, we focus primarily on the use of smart technology to monitor and assess human health and wellbeing.

In their assessment of self-tracking wearables, Crawford et al. (2015) claim that related technologies have a series of things in common. First, as purportedly radical technologies, they are associated with an ability to deliver 'precise and unambiguous physical assessments' (Crawford et al., 2015: 480). The ability to understand the real state of the body is in part a product of the non-subjective nature of the digital measuring devices (Crawford et al., 2015; see also Davidson, 2016). But it is also a feature of the epistemological logic of learning what is going on in one body by comparing it to a much larger pool of training data from other bodies. Second, smart body technology advocates tend to assume that human understanding and self-knowledge can be improved through quantitative measurement and external monitoring (Crawford et al., 2015). This assumption does, of course, reflect the broader epistemological orientation of the smart-tech society. Third, according to Crawford et al., smart body technologies assume a virtuous line of connection between data, bodies, and self-improvement (2015: 480). In these contexts, it is clear that smart body technologies speak to the heart of the wider issues that animate this volume, namely: How can smart digital devices change how people know themselves and the world around them? How can related devices change what we know? And to what extent does smart technology translate knowledge of self and others into action? In this chapter we position and analyse the various assumptions that are associated with smart body technologies. We do this initially through a consideration of existing theoretical perspectives and their implications, and then through an exploration of their practical use.

ANALYSING THE SMART BODY REVOLUTION

The Nature of Smart Body Technology

In this section we take a closer look at the empirical and epistemological nature of smart body technologies, and consider the existing interpretative frameworks that have been used to study them. An important starting point in these endeavours is a consideration of what makes wearable technologies smart. At the beginning of this volume we defined smart technology as 'any form of digital instrument that is able to learn from previous actions in order to optimise future ones' (Chapter 1, page 2). In this context, it is important to recognise the similarities and difference between smart body technology and the smart-tech systems we considered in Chapters 3 and 4. As with all smart-tech devices, the wearables associated with the smart body are characterised by an ability to offer continuous monitoring and automatic assessment of human bodies. They are also associated with claims of superior accuracy in their recording and assessment systems—particularly in comparison with more subjective human understandings of their bodies and health. Finally, and as with the examples of smart-tech systems we have previously considered, wearable smart tech offers feedback to users on their data performance and behavioural prompts to optimise future action. It is in relation to aspects of feedback, however, that smart body technology differs most from many other forms of smart tech.

Certain forms of smart technology relay information back to users: think here of a smart heating system or smartphone offering regular data updates on weekly usage rates and behavioural trends. It is, however, common for smart technology to internalise the lessons learned from human conduct (voice tone, dwell time, spelling errors, message response times, etc.). These data are then returned to users in the form of optimised options or behavioural prompts. In the case of most smart wearable technology, things are slightly different. Although user data are collected and aggregated by smart-tech companies (and used for secondary commercial services), generally the purpose of wearable technology is to provide an ongoing, live feed of bodily data back to users. It is in the context of this constant collection and relay of raw user data back to users that smart body technology is connected to the Quantified-Self Movement (Lupton, 2016).

The Quantified-Self Movement is an important point of reference for our discussion in this chapter. The term was first used in 2007 by *Wired* editors Gary Wolf and Kevin Kelly to describe an emerging group of people who were using wearable technology to record various aspects of their bodily existence. The unofficial motto of the movement is 'self-knowledge through numbers' (Wired Staff, 2009). As Lupton (2016) points out, the measuring and recording of aspects of one's self has ancient origins and has found more recent expression in the lifelogging movement. The movement's expansion is, of course, connected to the emergence of wearable smart tech. Lupton offers a helpful definition of quantified-self practices when she states that it entails:

> practices in which people knowingly and purposively collect information about themselves, which they then review and consider applying to the conduct of their lives. Self-tracking differs, therefore, from covert surveillance or means of collecting information on people that result in data sets to which the subjects of monitoring do not have access. (2016: 2)

While this is a helpful definition, it does reveal some of the definitional complexities that surround self-quantification and smart body technology. The use of smart body tech clearly involves a purposive decision to collect information about oneself, but it can also, at the very same time, involve covert forms of surveillance. Covert surveillance may involve the monitoring of activities of which users are not aware, or the gathering and analysis of aggregate data sets to which users do not have access.

In many ways this definitional tension speaks to the heart of why exploring smart body technologies is so important. The uncertain balance of benefits and costs associated with smart body tech is precisely what we are concerned with throughout this volume. It is in this context that we find another definition that Lupton deploys most helpfully. Lupton suggests that the Quantified-Self Movement is best understood as an '*ethos* and *apparatus of practices*' (2016: 3, emphasis in original). As an ethos, it reflects a desire to know oneself more completely through quantitative measurement. This is an ethos which is experimentally oriented and seeks to use changes in lifestyle, practice, and circumstance as an opportunity to monitor and measure bodily change (Wired Staff, 2009). The use of real-world experiments in this context reflects a kind of micro-scale reflection of the forms of continuous population-level experiments that are a key hallmark of the smart-tech society more generally. Also, Lupton (2016) notes that quantifying yourself is not always linked to self-improvement projects. It is in part about the 'lure of learning' (Wired Staff, 2009). But within the allure of learning remains the question of who precisely is learning the most about ourselves and the world around us on the basis of smart body technology. This question leads us to consider more deeply the 'apparatus of practices' that surrounds the Quantified-Self Movement, and who controls that apparatus.

In many ways it is helpful to think of the Quantified-Self Movement as a kind of socio-cultural response to the availability of smart body technology. While this chapter is not about the movement itself, it is interested in its underlying ethos and what this can reveal about the wider ambiguities associated with social responses to smart technology.

*

In the context of the smart body, then, smart technology operates in a very particular way. While many smart devices offer feedback that seeks to remove the cognitive load from users (in the context of automatically generated friend recommendations, search suggestions, film choices, or music playlists), smart wearables are often associated with making it easier for users to engage in cognitive deliberation. In this context, many users of smart-tech wearables do not use them because they remove the burden of having to reflect on data. Indeed, the motivation is quite the opposite. Wearable technologies are used precisely because they facilitate more data feedback to users about their body. There are some elements of smart-tech data processing/analysis going on in association with wearables (auto-generated trend patterns or daily comparison notifications clearly lighten the cognitive requirements of those keen to quantify themselves). But this happens alongside the relatively unfiltered presentation of user data to users. In this context, one of the key benefits of smart body tech derives from the fact that it automates what was previously a burdensome, analogue data-gathering and recording process. It does not, however, necessarily obscure valuable forms of self-knowledge.

In their analysis of the epistemological implications of wearable technology (particularly in relation to historical precursors of self-management), Crawford et al. (2015) acknowledge the tension, common with much smart tech, between enhanced self-awareness and exploitative forms of surveillance by others. But, in its constructions of particular ways of knowing oneself, wearable smart tech raises other questions about the epistemological assumptions and shortcomings of smart-tech knowledge regimes. In this context, our analysis of smart body tech is not primarily interested in the uneven systems of knowledge exchange it facilitates (with an inevitably uneven division of learning between individuals and Big Tech). Instead, we are concerned with the opportunities and limits of smart tech's ways of knowing.

The Opportunities of Smart Body: Empowering and Disruptive Technology

As we have previously mentioned, smart body technology carries with it the promise of an augmentation of human capacity to know and manage our bodies. As we outlined in the previous section, smart body augmentation can take two forms. First, it can facilitate the production of bodily data that act as a basis for conscious human reflection and deliberation on health and habits. Second, it enables the delivery of a series of behaviour-change techniques that target both conscious and unconscious forms of human motivation. In a study of 13 electronic activity monitors, Lyons et al. (2014) identified 35 behavioural techniques deployed across these devices. These behavioural techniques included goals and planning functions, feedback, social comparison, habit-formation techniques, and rewards and threats (Lyons et al., 2014). As we outlined in Chapter 4, smart technology clearly has the capacity for pernicious forms of behavioural manipulation. In the context of health-related smart body tech, however, it seems reasonable to assume that behavioural prompts will be in support of the actions desired by users (although as we consider in the following section, this may not always be the case).

While clearly seeking to augment the capacity of people to better understand and control their bodies and health, it is important to consider the efficacy of smart body technology. In their review of clinical trials of the use of wearable technology in weight loss programmes, Lyons et al. (2014) found no (statistically) significant evidence that the use of smart body tech in standard weight loss programmes changed participant outcomes. Indeed, Lyons et al.'s (2014) review suggested that face-to-face health interventions appeared to generally outperform eHealth interventions. This does raise the question of the overall health value of smart body technology. While less effective than in-person health interventions (including gym sessions or weight loss clinics), smart body technology would appear to offer a source of behavioural motivation that is convenient and accessible (Lyons et al., 2014). Smart body tech certainly provides a cheaper alternative to personal health care in the long term. It is perhaps the potential reach and low-cost nature of smart body technology—as opposed to its actual impact on individual health outcomes—which makes it attractive to public health authorities (Lyons et al., 2014).

Beyond the practical benefits of smart body technology, Lupton (2016) suggests that it could generate broader political opportunities. Lupton draws a line of comparison between Haraway's figure of the cyborg (see Chapter 2, this volume) and smart body tech. Lupton suggests that related technology could have positive disruptive implications. With the exception of smart-tech implants, wearable smart body technology arguably brings humans closest to Haraway's (1987) vision of the melding of biology and technology captured in the figure of the cyborg. If smart body technology is to be considered a manifestation of our cyborg future, its significance need not only rely on its efficacy. Haraway (1987) suggests that as an imaginative resource, the cyborg seeks to subvert established norms around the human body (including questions of race and gender) and facilitate new perspectives on what it is to be human. It seems to us that the proliferation of smart body technology offers the opportunity to generate a multiplicity of perspectives on the human body and broader human condition. Smart body technology could be thought of as a collective experiment in the study of human health and wellbeing that serves to (politically) disrupt established socio-biological norms. Lupton (2016), however, points out that many smart body technologies carry with them existing assumptions about what a normal functioning body is.

Internalised Surveillance and Body Rendition

There is, as you may expect, no shortage of critical interpretations of smart body technology and associated digital bio-monitoring practices. In the context of surveillance capitalism, Zuboff (2019) claims that smart body technology is another circuit in the ever-deepening commercialisation of human nature. According to Zuboff, the smart body reflects a form of internalised surveillance, whereby the digital technologies carried on human skin enable the rendition of bodies to Big Tech (see 242–55).[1] To support her claims, she reflects on a University of Toronto study of nine fitness trackers (see Zuboff, 2019: 249). This 2016 report was titled 'Every Step you Fake' and explored the privacy and security settings of prominent fitness trackers (Hilts et al., 2016). The study revealed that it was common practice for fitness trackers to store and share users' information with third parties and data analytics interests. It was also revealed that the personal information that was gathered and stored about users was not necessary for the tracker's operations (Hilts et al., 2016).

Lupton develops a more optimistic account of the human implications of smart body technology, but she does acknowledge the problematic surveillance implications of wearable technology. Lupton's account of forms of data betrayal echoes, but also extends, Zuboff's insights. According to Lupton, in

---

[1]   Zuboff's assertion that the emergence of the smart body is first and foremost about the rendition of ever more intimate details of human existence is supported by Apple's CEO Tom Cook. Cook described the Apple Watch as the 'most personal device' created by the company (Colt, 2014). Through haptic functions, smartwatches use touch to monitor what is happening in a wearer's body, and send prompts to the user via vibrations. Reflecting on Apple's official claims about their watches, Lupton notes that such technologies provide not only immediate feedback, but also more intimate forms of communication (see Lupton, 2016: 73).

addition to offering surveillance revenues for the corporate sector, wearable technologies also raise the prospect of new forms of governmental power, as states are able to better understand the behavioural patterns of their populations and instigate related behaviour-change initiatives (2016: 117). Lupton also emphasises the ways in which the data extracted by smart body technologies could be used to deepen existing forms of digital discrimination (2016: 119). With the data generated from smartwatches already being used to determine premiums on life and health insurance schemes, Lupton suggests that such data could determine access to credit, health care, and employment opportunities (2016: 117). In this context, it is not just that the extraction of personal data is used to identify existing forms of socio-economic discrimination, but to under-gird it. Finally, Lupton suggests that there is an increasing danger that smart body tech will transition from being an optional utility to being a required appendage to the human body. According to Lupton, the transition from wearable tech as a form of personal pedagogy and self-motivation to imposed self-tracking can already be seen in a series of existing initiatives, including the insurance systems and company wellness programmes (2016: 121–5).

There is a final critical perspective on smart body technologies that is important to highlight here. Zuboff argues that the origins of smart body technology can be traced back to the pioneering work of R. Stuart Mackay (Zuboff, 2019: 204–207). Mackay was instrumental in the development of telemetry (long-distance digital data transmission) as a technology for monitoring animal populations in the wild (Zuboff, 2019: 204–207). On one expedition to the Galapagos Islands, Mackay attached telemetric sensors to iguanas and got giant tortoises to swallow digital devices (see Zuboff, 2019: 205). Mackay's work would support attempts to better understand and protect endangered species. But according to Zuboff, it was also a troubling precursor to biomedical telemetry in humans. One of the key principles of Mackay's work was that to be scientifically reliable, animals carrying telemetric sensors should not be aware of their presence. Awareness of monitoring in this context would introduce a variable that could invalidate the data that telemetry could produce. According to Zuboff, 'Mackay's interventions enabled scientists to render animals as information even when they believed themselves to be free, wandering and resting, unaware of the incursion into their once mysterious landscapes' (2019: 205–206). Zuboff suggests that smart body technology effectively converts humans into animals in a biological study. Telemetric signals could be extracted from wearable technologies to reveal geolocational and bio-digital data about humans. When aggregated, these data could provide new insights into the nature of the human species 'in the wild', just as Mackay's work had done for other species on the Galapagos Islands. Most troubling for Zuboff, however, is the way in which smart body tech-nology operates at an unconscious level. As smart body technology becomes an accepted part of bodies, Zuboff suggests that we will become gradually unaware of its monitoring activities and reveal more aspects of our personal lives to surveillance capitalists.

Zuboff's analysis of biometrics and smart body technology is unyieldingly bleak. Given that smart body technology often seeks to inform users of their embodied actions, it is also clearly only part of the story. In the remainder of this section we introduce analyses that emphasise a more diverse and potentially productive set of capacities that are associated with wearable digital technologies.

Moral Epistemologies and Self-Knowledge

Perhaps the most interesting lines of inquiry concerning smart body technology are those that are concerned with the underlying, but often undocumented, effects that self-monitoring has on the nature and purpose of human self-knowledge. These lines of inquiry are perhaps expressed most clearly in Lupton's (2016) sociological analysis of self-tracking, but they are also evident in related forms of critical analysis (Crawford et al., 2015; Davidson, 2016). According to Lupton, smart body technologies reflect more than just the fortuitous coming together of smart tech and a desire to quantify ourselves. They also embody deeper socio-cultural trends that are synonymous with neoliberal values (Lupton, 2016: 139). These relate to the value that is placed in market-oriented societies on self-awareness and self-entrepreneurship. In such cultures self-care is seen as an individual responsibility, and successful self-care is a product of, and basis for, competitive advantage (Lupton, 2016: 139).[2] Lupton also claims that the rise of digital self-monitoring is the product of the privileging of particular scientific and objective ways of interpreting the human condition.

In the context of the broader mores of neoliberal societies, Lupton argues that smart body technology both reflects and supports particular moral epistemologies. Lupton states that in the world of digital self-tracking, 'selves and bodies are understood as atomised, shaped by personal life experiences and empowered to manipulate their destinies by acquiring self-knowledge and acting rationally upon that knowledge' (2016: 140). Notice here the parallels and distinctions with our previous discussion of the impacts of smart technology on human assumptions and norms. In Chapter 4 we charted the connections between smart technology and an understanding of humans as

---

[2]   This does not, of course, explain the uptake of self-monitoring in more authoritarian societies, or indeed the use of smart technologies to support more interventionist forms of government in personal life.

individualised behavioural units. This epistemology of the human condition as individual/behavioural disconnects people from the broader conditions that lie beyond individual choice but still determine life courses (such as educational opportunities, ethnicity, gender/transgender identity, class, etc.; Lupton, 2016: 140). But the behaviouralist ethos of the smart-technology society appears to be complicated somewhat in the context of self-monitoring. While certain forms of smart technology seek to compensate for the irrational shortcoming of humans, according to Lupton at least, digital self-tracking seeks to augment people's rational capacities (in this context at least, it is perhaps more neoliberal than *neuroliberal* (Whitehead et al., 2017). According to Crawford et al., the moral epistemologies associated with self-knowledge technologies operate in two ways: (1) to suggest a moral duty to know oneself; and (2) to assert that such knowledge is needed if effective self-improvements are to be made (2015: 486). In a fascinating comparison of weighing scales and wearable technology, Crawford et al. (2015) suggest that the moral connection between self-measurement and self-improvement can be traced back to 19th-century branches of liberalism. They discuss how weighing scales generated a popular acceptance of the connection between external measurement and personal wellbeing. They further chart how the transfer of the weighing scales from the clinic to public weighing scales, and eventually into the private home, was synonymous with an emphasis on the moral duty of self-monitoring as a pathway to personal betterment (Crawford et al., 2015: 483). The implications of Crawford et al.'s (2015) analysis are clear. Wearable smart body technologies are synonymous with the movement of health measurement from the collective sphere of health care and the doctor's office to the individual body. With this transformation comes a neoliberal moral assumption of a transfer of responsibility for health from the collective to the personal.

In drawing a line of comparison between the weighing scale and wearable tech, the work of Crawford et al. (2015) reveals that the changing moral epistemologies of the present may not be as novel as we assume. Wearable technologies do present opportunities for constant monitoring and automated data sharing that were not feasible when weighing scales first emerged. Crawford et al. (2015) do, however, emphasise the scientific similarities between the two technologies. The emergence of the weighing scale and smart body tech are synonymous with the valorisation of the precise measurement of human experience. This emphasis on precision seeks to militate against the vagaries of human subjective experience but can also result in a flattening of the variety of ways in which we experience and understand our own bodies (Crawford et al., 2015). This flattening or narrowing of experience is a theme that is explored in interesting ways by the geographer Joyce Davidson (2016).

Davidson recounts her experience of bodily quantification in relation to her engagement with medical professionals following a fall and suspected concussion, and pregnancy. In the context of the direct experience of separate attempts to diagnose a suspected concussion and assess the health of her twin babies, Davidson explores a tendency within modern health care to value what can be seen and measured over and above what can be felt. Davidson recounts her surprise that in the medical assessment of her health, how she felt was largely seen as irrelevant to the assessment of wellbeing. Echoing Cheney-Lippold's (2017) description of the *data self,* Davidson describes how she felt that her consultants were not dealing with her as a person, but as a *data double* (Davidson, 2016: 31). Davidson suggests that the medical focus on data generated a form of 'slippage … between observation and explanation', whereby only those things which leave a data trace are of elucidatory value (2016: 32). This led to a peculiar situation whereby a body that 'feels fine' is not to be ignored (as it historically would have been) but is to be continually monitored for measurable aberrations. In the era of the digitised body, then, it is just as much the healthy body which needs to be monitored as the unhealthy one!

Davidson draws on her experience of having her body digitised to question the value and assumptions of broader attempts to technologically monitor and quantify human health. She notices how smart body technology has been designed to generate 'useful emotions' such as guilt and shame in order to stimulate changes in people's health behaviours (2016: 32). For Davidson, however, the main emotional response she associated with monitoring was one of anxiety (2016: 32). Davidson's experience of digital body monitoring appears to reflect one manifestation of the broader rise of what Isin (2004) has termed neurotic citizens. In this context Davidson questions whether moves towards the quantified self are really about promoting self-reflection, as is claimed, or more centred on self-inspection (Davidson, 2016: 32). The distinction between self-reflection and self-inspection is an important one within the smart-tech society more generally. This distinction exposes a critical difference between the liberating potential of self-understanding and more restrictive forms of self-surveillance. Davidson also draws attention to the implication of bodily surveillance for questions of trust. The quantification of the body carries with it an imperative to trust the numbers (Davdison, 2016). By definition, trusting the numbers suggests that human opinion is to be trusted less. According to Davidson, the systems of trust associated with bodily quantification and digital learning have potentially broad social implications:

> I would suggest that, whereas typical trustful relations work to strengthen our sense of connection with other people, and our feeling of 'rightness' in place, trust in numbers functions largely to isolate us. Arithmetical trust subtracts us from potentially meaningful contexts and relations, other than, perhaps, competitive ones, where we evaluate how we measure up against others. Such metrics are masculinizing in that they fail to account for intersubjectivities, emotional relations and mutual

reliances. They operate in virtue of and intensify the illusion that selves operate atomistically, as singular, self-governing, systems. (2016: 32–3)

Here Davidson's analysis supports and extends Lupton's neoliberal critique of the Quantified-Self Movement. Lupton suggests that the quantification of the self reflects a broader system of self-care that is synonymous with neoliberal social norms. Davidson, however, argues that quantification not only focuses our attention on ourselves, but also generates an epistemological system that is actually antithetical to the construction of broader forms of social trust and collective support.

Although we are sympathetic towards the critical perspectives that Crawford et al. (2015) and Davidson (2016) develop, we are conscious of the potential limitations of their perspectives. We are mindful of the forms of social connection and care that can emerge within communities of quantification. We are also conscious of the fact that while smart body technology may narrow our epistemological vision, it does not necessarily preclude other ways of relating to and understanding our bodies. We explore these tensions in the ethnographic case study we outline in the remainder of this chapter.

QUANTIFYING THE SELF—AN AUTO-ETHNOGRAPHY

The remainder of this chapter focuses on an experiment that was carried out by one of the authors of this volume (Whitehead). The experiment is best described as an auto-ethnographic analysis of the use of wearable smart tech and integrated health apps. The trial was designed to explore the lived experience of using smart body technology. But it also sought to explore the processes associated with the production of a more quantified self. As we discussed earlier in this chapter, the Quantified-Self Movement is not synonymous with smart body technology. The movement, and its related practices, have, however, expanded greatly as a consequence of wearable digital technology. The process of self-quantification also embodies many of the underlying epistemological tendencies associated with the learning loops and feedback systems of smart body technology.

The experiment ran for a three-week period between 2 August and 21 August 2021. It involved the use of an Apple Series 3, 38 mm smartwatch, an Apple iPhone, and related health apps. The smartwatch was used to record steps, calories burned, sleep patterns, exercise, and heart rate. In addition, the Sleep++ and Waterlogged apps were used to record sleep and water intake, respectively. To facilitate reflection on the quantitative data that were gathered, a record was kept of the data for six key indicators (heart rate, exercise, sleep, calories burned, fluid intake, and steps) in a spreadsheet. During the 21 days, daily diary entries were made which explored: (1) observed changes in bodily activities; (2) the presence and impact of behavioural prompts; and (3) the overall experience of using smart body technology. As an auto-ethnography, this study sought to record personal reflections on the lived experience of using smart technology, which included feelings, anecdotes, and relevant stories. In this context, this short study was designed to reflect the techno-phenomeno-logical perspective favoured within this volume (see Chapter 2). The trial also utilised these reflections as a basis for considering more closely the opportu-nities, challenges, and problems associated with smart body technology. What follows in this chapter will inevitably have a slightly different feel to what has so far been written in this volume. It will take on a first-person account of the lived experience of using smart body technology. We hope that the change in writing style will not be too jarring. But we feel the short loss of narrative consistency is worth the sacrifice for the novel insights that auto-ethnography can bring.

Beginning Quantification—From the Qualitative to the Quantitative Self

Like all experiments, I suppose, the commencement of this trial felt arbitrary and a little unsettling. I had actually delayed the start of the trial to make sure that I would be able to devote enough time to it, but this felt like it was already introducing a form of bias. The first few days of smart body self-quantification felt strangely labour-intensive. I had to get used to starting my smartwatch's sleep-monitoring function before going to sleep; I had to work out how to effectively calculate the quantities of fluids I was consuming (how many milli-litres of fluids could my tea mug hold?); I had to figure out how best to extract bodily data from my smartphone (it turns out there was so much recorded data about my daily activities recorded on my iPhone's health app that finding the specific data points I was interested in was quite a challenge); I also had to get used to using the new apps I was utilising to record my sleep and fluid-intake data. In the early days of the trial it certainly did not feel like my smart body tech was making my life easier. Rather, I had the distinct sense that I was serving it. The need to record my fluid intake during the day felt like a par-ticular inconvenience, and I resented the extra complication it added to my already busy schedule. It was, however, the recording of my sleep that gave me the greatest pause for thought. In his analysis of capitalism's assault on sleep, Crary (2014) demonstrated how undermining sleep could become increasingly central to maintaining the profit margins of modern industrial society. I was also aware of the fact that streaming platforms, such as Netflix, saw the need for sleep as the enemy of their economic model. Thus, wearing my smartwatch while I slept felt like a significant moment—not so much a capitalist assault on my rest, but a surveillance capitalist attempt to extract profit out of the act of sleep itself.

After the first week of the trial, habituation meant that the concerns I had about monitoring my sleep started to fade. I also found that as I got used to the technology and apps I was using, things became less challenging and felt generally less intrusive. The emerging ease I had with smart body quantification is itself an interesting insight. It seemed to reflect a point when habituation meant that using the technology was no longer an inconvenience. Indeed, as this research diary entry revealed, I became increasingly aware of how difficult self-quantification and monitoring would be without smart-tech devices: 'Without automatic recording of much of my data, I think I would struggle to keep up the discipline for consistent self-monitoring. It just adds that extra level of complication and pressure into the day. Clearly, automatic data capture has made this process of self-knowledge production easier' (Self-Quantification Diary Entry).

At the same time, however, habituation also meant that I stopped critically questioning how and why I was using the technology. It is, perhaps, in moments like this that the convenience gains of smart technology may not necessarily be in sync with human empowerment. It does become easier to know yourself (at least in quantitative terms), but the ease of knowledge accumulation comes at the cost of moving the acts of self-monitoring into the unquestioned realms of the technology unconscious.

One of the most striking aspects of the early phases of digital self-monitoring is the emergence of the gap between the qualitative and quantitative self. The gap between the everyday lived experience of your body and its conversion to a measurable digital form echoes Cheney-Lippold's (2017) reflections on the data self. But in the work of Cheney-Lippold, the distinction between the data self and the analogue self is primarily expressed in terms of the formation of an alternate virtual self that is algorithmically constructed from the fragments of our existence that are recordable in digital form. In this context, Cheney-Lippold's data self is a digital projection of ourselves, but one that remains functionally separate, and often hidden, from our corporeal existence. Smart body technology collapses the gap between corporeal experience and digital monitoring. The quantified self is always already a qualitative self who cannot help but notice the disjuncture between the digital monitoring of bio-circadian rhythms and our direct sensory experience of our bodies.

My self-quantification diary indicates that I started to notice the distinctions between quantification and lived experience early in the trial. Here I reflect upon an emerging tendency to convert everyday experiences in quantitative outputs as I was experiencing them,

> Overall, I am noticing a conversion in my mind of pleasurable activities into quantitative outputs. I had a lovely sea swim with my daughters today, but my mind kept drifting on to its impacts on my calorie burn … What is this gap that emerges between measurement and experience, and does it matter? (Self-Quantification Diary Entry)

I noticed the benefits of converting qualitative experiences into quantitative form (particularly in relation to the comparative opportunities it provided), but remained suspicious of its overall implications. Here are two diary entries that reflect in slightly different ways upon the gap between the qualitative and quantitative self:

> I have noticed how I am beginning to convert mundane activities like mowing the lawn into quantitative exercise gains. There does feel like a cognitive shift as I move my everyday experiences into a quantifiable digital form. There are clearly simplifying gains to this. You are doing better or worse than the day before. But it does feel something of a barren form of existence. (Self-Quantification Diary Entry)

> While running today, I did start to reflect upon the distinction between the run of five miles and what it actually means. This felt like a functional run compared to the one the previous night. The quality of the run's difference could not, however, be conveyed in relation to a difference in their length. The fact that today's run was slower perhaps conveyed that it was tougher/less pleasant. But generally, quantification appears to open up a gap between representation and experience. I am consistently struck by how my experience of the thing I am monitoring is not encapsulated in the data. Despite the cleaving of data from experience, I constantly find myself trying to quantify my experience, while I am experiencing it. Perhaps this is why I stopped to meditate during a run this week—a search for a deeper form of experiential meaning in my bodily activities. (Self-Quantification Diary Entry)

Two issues emerge from these reflections. The first is a tendency while using smart body technology to give quantitative meaning to qualitative experience even before it takes numerical form (i.e. following data upload): I found myself thinking 'this activity is meaningful because it will produce quantitatively measurable results'. The second was my clear desire to try and pull back from the quantitative impulse in order to facilitate a more emotionally meaningful experience from a run (by incorporating an aspect of meditation into it).

Perhaps the most significant way I noticed the gap between a quantifying and enduring qualitative self emerged when quantitative measures did not directly tally with lived experience. Here are two diary entries that reveal this gap between experience and digital monitoring:

> [It] was intriguing given the gentle pace of my run today that I had my highest heart rate of the week. I checked back through the data and noticed that it was fairly early in my run. Is this a data error, or a weird increase in heart rate for no reason—is this hypochondria I sense? I did find myself going back through a longer backlog of data to give me peace of mind. (Self-Quantification Diary Entry)

> Again the gap between perception and data opened up today as my most restless night of sleep on record appeared to correspond to a good and long sleep. I wonder whether my attempts to sleep in a little added to [the measured] disruption. (Self-Quantification Diary Entry)

In the first case, my noticing of a high heart rate (with no corresponding increase in exercise intensity) led me to simultaneously doubt the technology and my own experience. In her analysis of self-quantification, Davidson notices how experts tend to stop trusting the individual as a reliable basis for insights into health and wellbeing. My experience of self-quantification appeared to lead to a more general sense of doubt as I was unsure whether to trust the data or myself. It is perhaps interesting that I used previous data records to reassure myself that I was OK. In the second instance, I became aware of the difference between what felt like a good night of sleep and what was actually measured as a restless one. What is most interesting about this example is that it led me to question what could technically be leading to this disjuncture (could staying in bed longer increase my positive experience of sleep but increase measures of disruption?). Ultimately, my feeling was that a motion sensor on a watch is a much less reliable indicator of the depth and quality of sleep than how I felt the following morning!

Comparison, Change, and Consistency

I thought initially that quantifying the self was about the production of discrete forms of self-knowledge. As soon as the trial commenced, however, I realised that the primary purpose of quantification is comparison. To choose quantification over experiential reflections reveals a preference for a form of data purpose that resides in aggregation and comparison. Although qualitative data allow for comparison, they do not require comparison to be meaningful. Without comparative frameworks, however, quantitative data are strangely meaningless. And so, in the first few days of the trial, the data I was accumulating about myself felt peculiarly worthless. Was that heart rate normal? What was an appropriate exercise level? What did a good night of sleep look like in quantitative terms? The problem was I had no frame of reference for my data.

The search for comparative meaning in the use of smart body data requires time and the accumulation of enough bodily information to base assessments upon. It also involves the establishing of, often, arbitrary goals against which performance can be gauged. In this context, it took several days for the processes of self-quantification to become meaningful. Once I had the capacity for data comparison, it remained unclear (at least to me) what the actual purpose of comparison was. Sometimes it appeared to be about establishing forms of behavioural consistency. At other times it appeared to be about generating deliberate changes in bodily outputs. The desire for consistency was expressed most clearly in the quantification of sleep. The importance of sleep consistency was emphasised by the Bedtime Consistency Score on the Sleep++ app I was using. This score measured consistency in the time you go to bed each night. The biological purpose of this consistency score appears to relate to its role in helping establish constancy in the circadian rhythms of the user, which will be mirrored in the consistency of subsequent sleep. Here are my diary reflections when I first discovered the Bedtime Consistency Score:

> The Bedtime Consistency Score made me think of the ways in which self-quantification does not just focus on the maximization of certain data inputs and the minimisation of others, but also the establishment of consistency in some. So, while the motivational trend line in the data is generally up, in some instances it can also be about consistency. I am not sure why consistency of sleep is important, but the score needs a three-night average to establish and use a ±10 minutes scale. It is clear that consistency is valuable for habit formation though. (Self-Quantification Diary Entry)

Interestingly, the very act of quantifying sleep was what first led me to pursue a consistency in my bedtime (this was before I was even aware of the Bedtime Consistency Score). The only way I could achieve consistency in my sleep length was to go to bed at the same time every night (I did not have the luxury of staying in bed longer in the morning, that was for sure). My smartwatch provided me with a behavioural prompt every night by playing a gentle lullaby ringtone. I found this prompt to go to sleep one of the most frustrating and restrictive aspects of the whole trial. I desired greater flexibility in when I chose to go to bed. The quantitative drive for consistency did, however, prove to be a significant motivational drive.

The pursuit of consistency in sleep led to me establishing a much more regular sleeping routine and ultimately to me sleeping longer. On these terms, at least, it appears that the use of smart body technology and self-quantification was beneficial to me. I certainly started to feel like I had more energy during the day. One interesting aspect of quantified performance, however, was noticing the factors that disrupted consistency. Here are two diary entries that reflected on this issue:

> Achieving consistency in self-quantification requires routine. It is interesting to see how, over the weeks, keeping track of my bio data has generated a form of patterning in my actions, including drinking water and sleeping. When that pattern is disrupted, as it was today, it is difficult to maintain those routines. To these ends, the focus of self-quantification on my bodily activities has served to remind me of the social and habitual context of my body and the influence of these contexts on what my body does. (Self-Quantification Diary Entry)
>
> [I had a] realisation about how bodily habits are tied to working habits. It is interesting to think about how self-care is graphed onto other work-based prompts—the need to wake up at a certain time, the need to have a break, etc. (Self-Quantification Diary Entry)

The second quote reflects on the most disruptive threat to consistency: the weekend. In the absence of the natural routine of the working day, it was particularly difficult to structure sleep and fluid intake. Even though I technically had more time to achieve my quantification goals on the weekends, I was generally less able to attain them. The first quote is taken from the day I had to go to sleep much earlier than usual in order to be able to get up in time for a meeting I had to travel to. These two disruptive moments made me realise how consistency in my smart body performance was reliant on a degree of consistency within the social context of my life. By definition, then, self-quantification is not only about attempting to control what a body does, it must also seek to control key aspects of the social life of that body.

It is tempting to equate the impacts of self-quantification with the capitalist colonisation of human time described in Crary's (2014) aforementioned book, *24/7*. In this volume, Crary describes the emerging attempts that have been made to overcome the limits placed on economic production and consumption by the natural rhythms of human life (and in particular, sleep). According to Crary, modern innovation in production and consumption associated with a 24/7 culture 'disavows its relation to the rhythmic and periodic textures of human life' (2014: 9). For Crary, capitalism's assault on human nature serves to homogenise time and remove the inconvenience of interludes, cycles, and seasons. In some ways, smart body technology and self-quantification could be seen as part of this homogenising temporal impact of capitalist 24/7 culture. It could be interpreted as an embodied attempt to produce greater consistency in human activities that overrides nuisances such as the weekend. But this account is, at best, partial. The temporal consistencies associated with smart body tech are not necessarily about the economic performances of production and consumption (although they can be—see Chapter 6). Rather, they are focused on the regularities of wellbeing- and health-enhancing practices: 24/7 culture is about less sleep, self-quantification is about more (or at least more consistency). Additionally, in my experience, the functional effectiveness of self-quantification appears to depend on the use of human rhythms and habits as natural interludes for the recording of quantified data. Thus, while self-quantification encourages consistency, it would appear to benefit from human cycles (the need for water, rest, toilet breaks) as natural punctuation marks for measurement and quantification. If industrial capitalism seeks to eviscerate human nature, surveillance capitalism seems, at least in part, to work with human nature. The smart-tech society appears to form relations with naturally occurring bodily patterns as a basis for novel forms of measurement and ultimate exploitation.

The other form of comparative goal that characterises smart body tech is the measurement of change. The detecting of change is possibly the most important dimension of self-quantification. Change is what triggers behavioural prompts to action; change is what instigates the search for interpretation; change felt to me like the driving force behind self-quantification. In the early days of wearing smart body tech, the lack of historical data made feeling a meaningful sense of change difficult. But once sufficient data were available, fluctuations in bodily performance became a prime focus of self-quantification. The following section considers how the detection of change (perhaps relative levels of inactivity or overachieving in exercise) becomes a basis for smart-tech prompts for behaviour change. But at this point I want to focus on the measurement and assessment of change in broader terms.

The smart body tech I used measures and assesses change in two ways. First was the use of average measures of steps, exercise levels, and calories burned. These average measures were used to provide regular updates during the day pertaining to how I was performing in relative terms to my average. Second was the use of standard units of aggregate comparison, primarily days and weeks, to assess performance change. During my experiment in self-quantification I became increasingly aware of the arbitrary nature of daily and weekly performance as units of assessment. Here I reflect on this issue in relation to the allocation of performance awards by my smartwatch:

> It is interesting how [digital] awards are given out on a weekly basis. When it comes to quantification, days and weeks, and hours, become important units of measurement of activity. I always feel sleep disrupts these measures as it does not really fit into one day. But, to quantify, you need blocks of time, however arbitrary, because you need discontinuity in measurement and comparison. I don't feel that the weeks really reflect my experience of biological life and activity, though it is another way in which experience and measurement appear to diverge. (Self-Quantification Diary Entry)

The use of days to assess performance became a source of frustration because poor performance on the previous day was a basis for an easy overachievement in performance the following day, while a high-performing day brought comparative pressure to the following day. Weekly comparative measures also felt arbitrary as weekly fluctuations in performance tended to mask broader trends. It was also unclear when you should measure a week. Should it be the last seven days (and thus a constantly moving unit), or Monday to Sunday? The use of 'yesterday' and 'last week' as comparative measures of change are clearly attempts to make bodily performance data meaningful on human terms. But these units of assessment oversimplify changes in bodily performance, in my experience at least, and were as much sources of frustration as motivation. But—and this is important—they were still sources of motivation in their own way (I will discuss this in the following section).

*

The monitoring and assessment of bodily change and constancy raised a consistent question in my mind throughout the trial: Should smart body monitoring be an objective scientific study of the body, studying its activities in a relatively indifferent and arbitrary way in order to understand its natural state? Or should self-quantification be the basis for deliberate change in embodied practice? As I completed the first week of my trial, I noted the following observation in my research diary: '[I am] conscious of whether I am trying to experimentally monitor myself now or use measures to change myself. There is at the very least a transition in my mind between seeing what I am, to exploring who I could be!' (Self-Quantification Diary Entry).

There is, as you might expect, no rulebook on this issue. Smart body tech and the associated Self-Quantification Movement are open-ended collective experiments that can be taken in a variety of directions and used for varied purposes. While I was perhaps unclear precisely what the purpose of the experiment I was conducting was at the beginning, it very quickly took on its own logic of personal improvement. Here is the very first entry into my diary:

> I noticed a definite heightened awareness of my body [today], what it was doing, and what was going into it. I also felt a pressure to remember to record, but not a clear sense of why I was doing this. There was a sense of satisfaction when reviewing the previous day's data—a sense of empowerment and control perhaps? When I noticed the potential for daily and weekly comparisons, I could also sense a competitive element rising around this. Perhaps the first day felt like a controlled experiment, but the future felt like something I wanted to enhance off this baseline. (Day 1—Self-Quantification Diary Entry, 2 August 2021)

While the experimental purpose of smart body tech may be indeterminate and subject to change, what is beyond doubt is that it is experimental in its ethos.

As my own trial with self-quantification evolved, I became aware of the multiple and overlapping forms of experimentations that appear to infuse smart body technology. First, there is the aforementioned monitoring of the body in its natural form. Second, is the opportunity to induce continuous experiments to see how the performance of that body could be changed or made more consistent. In its early days, the Self-Quantification Movement emphasised how everyday life provides an almost infinite range of experimental opportunities to test and monitor bodily change (Wired Staff, 2009). Third, however, is the meta experimentation that informs smart body technology. This is the type of experimentation that feeds the architectures of surveillance capitalism, wherein bodily responses to experimental prompts generated by smart tech are monitored and assessed at the level of entire populations. The forms of large-scale experiments associated with surveillance capitalism stretch the bounds and potentials of self-quantification experimentation. Within the architectures of surveillance capitalism the opportunities for experimentation far exceed those envisaged by the Self-Quantification Movement. Surveillance capitalists have the potential to produce an infinite loop of simultaneous experiments that can be tweaked and repeated at almost no cost. The overlapping and varied forms and scales of experimentation associated with smart body technology are a reminder of its ambiguous implications for humanity. At one level, it provides the opportunity for the instigation of potentially empowering forms of experimental self-learning. Yet simultaneously it can involve the disempowering subjugation of humans to the unseen and unknown experiments of others.

Behavioural Goals and Prompts

In light of the way in which my trial ultimately became a quest for self-improvement, the final aspect I want to focus on is the issue of behaviour. As we discussed in the previous chapter, understanding and modifying human behaviour is a central component of the smart-tech society. The connections between smart body technology and behaviour change can be understood in at least three ways. The first is the use of technology as a basis for producing an historical record of comparison against which future behaviours can be measured (see previous section). Second is the use of devices to set and monitor goals, which essentially project desired behavioural achievement into the future. Third is the use of smart devices to provide prompts and nudges to action. In this section I focus on goal setting and smart prompts/nudges.

During the trial I set personal daily goals for water consumption, number of steps, calories burned, standing time, and amount of exercise. The Waterlogged app I used displayed my goal as a slowly filling water bottle (with the promise that if I upgraded to the premium account I would be five times more likely to reach my goal—an offer I respectfully declined). My Apple Series 3, 38 mm smartwatch displayed my goals as a series of concentric rings that would slowly grow during the day. Behavioural actuation is pursued by smartwatches through a series of measures. First are reminders of when daily activity appears to be down compared to the previous day. If this was the case, I received a haptic buzz and a short message to encourage me to either stand up or check my rings for exercise levels. Second is the use of awards. When each of my goals was achieved I received a congratulatory message, and when all three rings were closed, Apple offered a rather satisfying display as my rings glowing and sparkling like a firework display (Figure 5.1). Apple also rewarded me for longer-term achievements: perhaps the completion of a perfect exercise week, or achieving my longest streak of closed rings. These awards come in the form of digital trophies that have a personalised engraving on the back.

Source: Authors' collection.

*Figure 5.1   Apple Watch sends its congratulations after I close my
                 exercise ring*

Although smart behavioural prompts and rewards did serve to motivate me, as the trial continued I became increasingly conscious of the arbitrariness of the goals I had set. Here is one entry from my research diary on this theme:

As the days go on I become increasingly aware of the arbitrariness of most of the goals I am setting for myself, or that are being smartly set for me. The benefit of eight hours of sleep seems well established, but why do I need to walk 10 000 steps, or stand up during 12 hours of a day, or burn a specific threshold of calories? The comparison offered by Apple of the previous day's activities seems particularly arbitrary. Having completed 10 000 steps today I decided to explore the origins of that seemingly universal goal. Turns out it originates from a Japanese marketing campaign for a pedometer that was introduced before the 1964 Tokyo Olympic Games. They used the term Manpo-kei (10 000 man): po for steps and kei for measure. The goal was successful as a campaign. Research now appears to show that 7500 steps is enough to get the necessary mortality benefits. I did find reference to the benefits of 150 minutes of moderate exercise per week, which could be useful in exploring longer-term trends. (Self-Quantification Diary Entry)

Motivational goals can be adjusted for the user, and my watch even suggested lowering them if I consistently fail to reach them. This system of adjustable goals appears to reflect an interesting interaction between smartwatches and users in the pursuit of an optimal level of motivation, achievement, and reward. However, my research on the origin of many default goals made me question the potential disjuncture between what works as a smart motivational goal (the symbolism of reaching 10 000 steps) and what is actually needed to achieve desired health benefits. This is, perhaps, an overlooked aspect of self-quantification: the way in which the quantification of experience intro-duces the prioritisation of goals around seemingly significant round numbers.

The maintenance of streaks provided a surprisingly powerful motivation for meeting my bodily goals. Before beginning the trial, I hadn't paid a great deal of attention to the streaks that my smartwatch kept referring to. But engaging in self-quantification inevitably meant that I was giving more attention to the consistency of my goal achievements. Interestingly, I did begin to speculate on the way in which streak maintenance appeared to tap into important motiva-tional forces that have been identified within the behavioural sciences:

> It is always nice to get a streak award. A lot of the behavioural prompts used by Apple Fitness appear to focus on streaks. The consistent meeting of goals and the keeping of a streak provides a fairly powerful motivation. To break the streak seems to undermine your own achievement and clearly taps in to psychologies of loss aversion—I am much more concerned about keeping a streak going (although not that concerned, I must confess) than establishing a streak. (Self-Quantification Diary Entry)

> I received an engraved shield for a perfect week of standing up. While superficial in some ways, the shield does feel tangible and significant. You can spin it around, and on its flip side my name is engraved with the date. It shouldn't matter, but I quite like the shield and the trophy room it goes in. (Self-Quantification Diary Entry)

The notion of perfection is a common discourse within the behavioural prompts and rewards associated with my Apple devices. As a goal, the very notion of perfection resonates with the numerical orthodoxies of self-quantification and digital body monitoring:

> There is a common refrain to perfection in the prompts I receive. The perfect exercise week, the perfect standing week. This idea of perfection suggests a consistency of performance which seems to be a common goal in self-quantification. The idea of perfection seems to me to be well aligned to the totalising measures of the quantified self. It does seem, however, to belie the inevitable imperfections of the human body and will. (Self-Quantification Diary Entry)

The goal of perfection reflects a novel aspect of the moral epistemologies associated with self-quantification identified by Lupton, Davidson, and Crawford et al. The attaining of perfection in personal performance not only reinforces the neoliberal orthodoxy of self-responsibility, but also raises the competitive bar. Improvement in the numerical performance of your body is no longer enough: smart body technologies reveal the potential for perfectibility and establish the motivational context accordingly. As my reflection above reveals, I understood the simplifying and unrealistic logics associated with quantitative claims to perfection. But as with so many of the behavioural strategies associated with smart technology, a realisation of the absurdity of consistently attaining perfection did not mean I didn't try. As I was told I had achieved perfection, I wanted more of it. As an achievement it seemed to be the only one that had unimpeachable value. In this context, closing the gap between reality and perfection appears to be another one of the human consequences of the smart-tech society—not so much smart tech augmenting our abilities, but raising the bar on our performance expectations.

In addition to target-setting and rewards, my smartwatch also deployed a series of more immediate behavioural prompts to action. As previously stated, these prompts came in the form of haptic buzzes, motivational messages, and reminders. One of the most common messages I received was one that encouraged me to stand up. I found that this reminder was pretty effective at inducing a sense of guilt and was generally successful in getting me to leave my computer to make a drink or have a wander around the house. Another common prompt was one to progress my exercise ring:

> I noticed the prompt today to exercise because I had done less than I normally do. Smart tech appears to like to use diurnal comparison to suggest action. This did feel quite a blunt tool yesterday. I had delayed my exercise yesterday because of various logistical issues. It seems like smart tech is looking for any break with the norm as a basis for intervention. (Self-Quantification Diary Entry)

The use of daily comparison as a basis for exercise prompts was enabled by processes of digital learning, but it seemed incredibly simplistic and did lead me to question the smartness of the technology. Perhaps this intervention is not a reflection on the actual sophistication of the technology, but just like messaging a friend, there has to be some premise, however arbitrary, for a behavioural intervention.

Shorter-term behavioural prompts clearly reflect the forms of micronudging we discussed in Chapter 4. Although the power of these forms of behavioural intervention can be easily dismissed, I actually found them to be fairly significant forces in shaping my daily routine. Drawing undoubtedly on behavioural and psychological insights, they seemed to strike an interesting balance between competitive encouragement and guilt. Although I found that I paid less attention to the prompts as the trial proceeded, I noticed how I had internalised their meanings. I could quickly interpret a haptic buzz as an indicator of whether I should stand up or needed to get out and do some exercise. This shift from the conscious acknowledgement of a behavioural prompt to a more automatic understanding of its intent seemed like a potentially significant impact of smart body tech. I may not, like Pavlov's dog, necessarily respond to the behavioural prompt, but I had been programmed to know what it was demanding from me.

CONCLUSION

It is highly ironic, but as I look back at the quantitative data and my qualitative reflections on my self-quantification trial, the experiment appears to have been a quantitative failure but qualitative success. At a quantitative level, my use of smart body tech and self-quantification did not lead to a discernible improvement in my daily metrics. In qualitative terms, however, I did experience discernible benefits in the process. The following research diary entries reflect on these generalised benefits:

> I am starting to feel what can be best described as pride and satisfaction in my morning data entry. I look forward to seeing how things compare, and I have a sense that the figures show I am generally leading a healthier lifestyle. The improvement in sleeping habits is particularly encouraging and comforting. (Self-Quantification Diary Entry)

> I do find myself taking some pride in my self-monitoring activity. Yes, it feels a little narcissistic, but it does generate a sense of taking control, however false this sense might be. (Self-Quantification Diary Entry)

My satisfaction in self-monitoring derived from a sense of inquiry into myself and my habits. While it did not necessarily mean that during the time of the trial I was any better able to control these practices, it did afford me a sense of control. In contrast to Davidson's suggestions, I did find self-quantification to be more about self-reflection than self-inspection (2016: 32). Despite generally feeling good about the whole process of self-quantification, I remained unsure about precisely what its beneficial purpose was. Here is a diary entry from midway through the trial which reflects on this point:

> I do find myself asking the overall question of 'what impact is this actually having on me—will I experience a meta effect in improved well-being?' I know well-being is much more complex and socially determined than that, but I can't help but wonder that if the physical building blocks are there, gains should be easier to activate. But then, is the purpose of this happiness? Risk management? Control? Pedanticism?

The reasons for, and experience of, digital self-quantification will obviously vary greatly, and the nature of my own response should perhaps not be valorised too greatly. However, there was one insight that emerged from the trial that I do think has broader implications for discussions of the smart-tech society. In becoming deliberately entwined in the learning loops and behavioural actuations of a collection of smart technologies, I started to question more deeply what it was that could be considered to be smart within these technologies. This extended diary refection provides insight into the nature of my question:

> The buzz on my wrist usually means one of two things: I am doing well (on average) or slipping behind. I kind of know which one it is without having to look at the actual message. The buzz is a convenient but fairly crude form of communication and it made me think about what is actually smart about my watch. The messages are personalised, but I gave it my name. The targets are monitored and recommended, but they are mine. The smartwatch appears to operate at a fairly basic level, taking up the cognitive load of remembering to encourage me to stand up because I will forget. But this seems like a very rudimentary form of 'intelligence'. It is not that smart tech cannot be cleverer than this, but many interfaces give us relatively little back—perhaps the smart insights are on the other side of the surveillance capitalism interface. (Self-Quantification Diary Entry)

It is by no means a new insight to claim that smart tech is primarily smart for its Big-Tech architects. Zuboff (2019) actually suggests that if a device is described as smart, this primarily means that the prefix can be replaced with surveillance. But if smart tech involves a trade-off between convenience and utility, and surveillance and manipulation, it seems important to pay attention to precisely what the utilities we are getting from smart tech are. In these terms, it may not be that smart tech is inherently good or bad, but that perhaps what I gained from using smart body technology in the short term is outweighed by what I may have unknowingly lost in the longer term!

6.   Smart working and the corporation

We begin this chapter by reflecting on an extended quote. This is Alex Stamos, former Chief Security Officer (CSO) at Facebook, speaking in 2019:

> The major tech companies are all acting in a quasi-governmental manner. When I was a CSO of Facebook I had an intelligence team. I had a team of people whose entire job was to track the actions of state governments and their activities online and then to intercede to protect the citizens of other governments … I had a child safety team. I had a counterterrorism team. These are governmental responsibilities that have been taken by the companies by the fact that they own the platform, they own where the data is, and they have access to data and resources that the public sector does not. The companies all have people who decide what is acceptable political speech, the people that decide what is acceptable advertising standards for people to run ads and democratic elections. These are government decisions that, generally, are being made privately, they have effectively speech police … So the companies are acting like governments but they don't have the legitimacy of governments. They don't have the transparency, they've never been elected, people choose to use their products, but they can become so powerful that they are acting at the same level as a government from a power perspective, where people can't really choose to be free of the indirect impacts of that platform; and, that causes a lot of problems. (Stamos, 2019)

We start with this quote because it offers an arresting segue into the emerging implications of smart technology for corporations, their employees, and their relations with wider society. Stamos's statement reveals the impacts that smart technology is having on the corporate form of Big-Tech companies themselves. It appears that the nature of certain companies directly involved with smart technology means that they now have the capacity and (at least assumed) responsibility to take a more active governmental role in social life.

In this chapter we explore the relationship between smart technology and corporations in two main ways. First, we consider how smart technology is used by corporations (within and beyond the Big-Tech sphere) to augment working life. Smart technology is now embedded within various aspects of working life. It is used by corporations to recruit talent. Related technology is also facilitating the production of quantified workplaces, wherein more and more of our working life becomes transparent and comparable. The Bank of America has, for example, recently deployed smart badge technology to monitor the movements of employees (Ollila, 2018). It is claimed by industry insiders that such smart badge technologies (which can incorporate Bluetooth proximity measures, microphones, and accelerometers) are like 'Fitbits for your career' (Heath, 2016).

The move towards the quantified workplace has, of course, been greatly accelerated by the COVID-19 pandemic as more and more of our working interactions occur in digital and, therefore, monitorable form. The ability of smart technology to learn more about how we work, and to augment the labour process, raises issues beyond those previously discussed in relation to the domestic use of smart technology. Unlike in the domestic sphere, the emergence of smart workplaces often sees people having far less choice about the nature of their engagement with smart technology. This raises important ethical issues. The presence of smart technology in the workplace also seems to be transforming key aspects of the labour process. It appears that smart technology is challenging many of the informal practices and tacit assumptions that have historically characterised the workplace and notions of expertise (Pasquale, 2020). At the same time, it seems that, rather than making working life easier for humans, working alongside smart tech means that people are increasingly expected to keep up with smart tech in the workplace.

Our second line of inquiry in this chapter builds more directly on the reflections of Alex Stamos. In recent years it has become apparent that Big-Tech companies have been drawn into the realms of government. The governmental role of smart-tech companies is evident in the acts of Twitter determining who has the right to public speech and who should be de-platformed. It can also be discerned within Facebook's attempts to promote voter turnout in elections. It has also been evident in Big-Tech's involvement in the governance of the COVID-19 pandemic. Many of these governmental duties are taken on willingly by Big-Tech, as they support and enhance their commercial goals. Others are more emergent qualities of the power, responsibilities, and social problems associated with the apparatus of the smart-tech society. It is important to note that we see these activities as being related, but ultimately distinct, from Big-Tech's involvement in the construction of smart states (see Chapter 7, this volume). These are activities that are not undertaken only to support the actions of government but involve smart-tech companies becoming governmental in their own right. In this chapter we consider the implications of Big-Tech being drawn into governmental activities (see Bartlett, 2018). Reflecting on case studies of the smart workplace in action and drawing on interviews with corporate insiders, this chapter considers the ways in which smart tech is transforming working life and changing the relationships between corporations and the public.

WORKING IN THE SMART CORPORATION

In this section we explore some of the implications of smart technology's deployment in the workplace. Our focus here is on the particular interactions between smart tech and human labour, broadly defined. Considering the human–smart-tech interface in this context is significant because it draws attention to the role of labour within human life (our work is obviously central to our sense of purpose, identity, wellbeing, and often dissatisfaction). Our analysis in this section also raises questions concerning which aspects of economic activity are likely to remain human-oriented, which can be replaced by smart technology, and which are likely to see a combination of people and smart tech.

A New Division of Labour: The Uneven Impacts of Smart Technology on Workplaces

The impacts of smart technology on working life are likely to be broad and varied. What is becoming increasingly clear is that there are few aspects of our relationship with our workplace that are likely to remain untouched by smart tech. From AI systems analysing our CVs and application letters, to wearable smart tech monitoring our workplace performance and deciding that we may be surplus to requirement, smart tech will shape our working lives and opportunities in myriad ways. However, while smart technology is likely to impact all workplaces to some extent, it is already clear that it will have a highly uneven influence on different sectors of the economy. The possible impacts of smart technology on workplaces can be broadly placed into three categories. At the most troubling end of the spectrum is the use of smart technology as a knowledge-extraction and labour-substitution tool. In this context, smart technology can learn from the operations of existing experts and professionals in order to optimise performance, before eventually replacing the humans from whom it has learned. Perhaps a more common impact is likely to be what could be termed smart-tech augmentation. Augmentation involves the establishment of working partnerships between humans and smart technology. This may involve a reduction in the number of people needed in the workplace, but not the total replacement of human labour by smart machines or AI. The final potential impact relates to a more general form of smart-tech monitoring and feedback in working environments. Such smart-workplace initiatives can inform knowledge-extraction and substitution systems as well as augmentation programmes.

However we characterise the impacts of smart technology on our working lives, its influence appears to be accelerating. The World Economic Forum

(2020a) suggest that working practices are currently subject to a double disruption involving the combined impacts of smart technology development and deployment, and the COVID-19 pandemic (see also Wallace-Stephen and Morgante, 2020). In the context of this double disruption, the World Economic Forum has conducted research with business leaders from around the world and developed estimates of the likely impacts of smart tech on different economic activities. Related research indicates that 45 per cent of businesses are planning to reduce the size of their workforce due to so-called 'technological integration'[1] (World Economic Forum, 2020a: 5). But this threat of replacement is complicated by the fact that 35 per cent of employers expect to expand their workforce due to technological change. The World Economic Forum also estimate that by 2025, although technological change may have resulted in the 'displacement' of 85 million jobs, '97 million new roles *may* emerge that are more adapted to the new division of labour between humans, machines and algorithms' (2020a: 5, emphasis added). Although the slightly menacing reference here to more *adapted* roles may give us pause as to what the smart-tech workplace may look like, business leaders are clearly not planning for a post-work world just yet. According to the Bank of England (2021), the jobs that are most likely to see a direct transfer of labour from humans to technology are in the fields of telemarketing, inspectors, electronic equipment assemblage, and drivers. But it is unlikely that dentists, air traffic controllers, and mental health professionals are going to be replaced by smart technology any time soon (Bank of England, 2021). It is clear that many lower-skilled jobs are at greater risk of replacement than professional careers, but this relationship is a complex one. The work of atmospheric and space scientists and accountants is more likely to be replaced by smart tech than that of event planners, for example (Bank of England, 2021).

What we can discern from the estimates of the World Economic Forum and the Bank of England is that the smart workplace will not simply be a place with fewer people in it. So-called dark factories, with no workers, may thus become more common in the future, but they are likely to remain the exception and not the rule. In certain sectors there will be less need for human labour, but in others the need for existing human expertise will endure. Reflecting on Tesla's failed attempts to fully automate its manufacturing process, Darling

---

[1]   We note that 'technological integration' does not just mean the use of smart technology. It may also relate to the use of robotics and automated systems that do not have smart capabilities.

conveys one of the reasons that human labour will continue to hold value in the smart-tech workplace,

> the robots, while able to work consistently and precisely, weren't able to recognise the litany of minor defects than can happen during the manufacturing process— slightly crooked parts, for example—leading to problems down the line. Human workers have the flexibility needed to recognize and correct unexpected errors in the assembly process, which is particularly crucial during the final assembly of a car. (2021: 13)

While smart technology may be able to offer consistency and precision, it appears that it is, as yet, unable to replicate humans' ability to flexibly adapt to the unexpected outcomes of complex processes.

In other smart-tech workplace contexts it is claimed that we should expect the emergence of new areas of human expertise and the creation of novel job opportunities. Business leaders anticipate that approximately 50 per cent of their current workforce will require reskilling (World Economic Forum, 2020a: 6). This mass reskilling will, in part, be a response to the internal redeployment of workers into new roles, but it will also involve those retaining their current roles adapting to technological augmentation in the workplace. The Bank of England suggest that the smart-tech world will be characterised by a shifting emphasis on the skills that workers will be expected to hold. Technological skills will be required to work within the increasingly digital spaces of the future. As previously mentioned, higher thinking and reasoning skills will also be valued in order to be able to problem solve in increasingly complex and multidimensional work contexts. The Bank of England also suggest that using social and emotional skills will be an increasingly valued part of human labour. It is, of course, precisely these social and emotional skills that smart technology finds hardest to replicate (Pasquale, 2020: 33–59).

The loss of a job, or reskilling within a career, will clearly be a significant part of the human experience in the smart-tech society. We do not wish to downplay the significance of the human experiences associated with the labour market displacements that smart technology will likely cause, or their uneven impacts on already disadvantaged groups. But in this chapter we focus on the implications of smart tech for those who will remain in the workplaces of the future. Initially we reflect on insights into what it is like to work in an existing smart workplace. We then consider some more normative issues concerning how best to construct the workplace of the future to serve human needs.

Technological Fulfilment: Working in the Ultimate Smart-Tech Firm

Amazon is an obvious, but also problematic, choice for a smart-tech workplace case study. Let's deal with the obvious first. From its fulfilment centre warehouses to its webservices and stores, Amazon is the living embodiment of the smart-tech firm. It has used smart tech to optimise all aspects of its economic processes. Its economic rationale is, perhaps, best summarised as an attempt to use smart technology to make commercial purchases as frictionless as possible (Galloway, 2017: 23). From algorithms anticipating what consumers need to the use of smart warehouses and logistics systems, Amazon's success can be directly connected to its unerring ability to digitally learn about the consumers and workers it engages with. A sense of the level of success Amazon has had in its use of smart-tech methods is perhaps captured best by the fact that Amazon (founded in 1994) had achieved $120 billion in revenue by 2016 (Galloway, 2017: 17). In other words, Amazon's revenues in the first 22 years of its life vastly outstripped the total income of Walmart (not exactly an unsuccessful company), which recorded revenues of $112 billion in the first 35 years of its existence (Galloway, 2017: 17).

The problematic aspect of considering Amazon is that its business model and the way it treats certain groups of its employees have proven to be controversial (see Pound, 2020). The controversial dimensions of Amazon's smart-business model means that analysing it with anything but predetermined assumptions is very difficult. The arguments surrounding Amazon's operations have also resulted in the company becoming defensive about its working practices and taking punitive action against whistle-blowers. Thus, finding out what working for Amazon is actually like can be challenging and tends to be divided between anodyne corporate messaging and the more dramatic accounts of whistle-blowers (Bray, 2020).

Notwithstanding these challenges, we feel that there are good reasons to focus our analysis on Amazon. First, with a workforce of approximately 850 000 workers, Amazon provides advanced insight into what happens when smart-tech working practices are scaled up significantly to large populations of workers. Second, as a leader in the field of smart-tech working, Amazon offers a glimpse of what the future may look like for many workers. While exploring the broad experiential implications of Amazon's working practices, in this section we focus specifically on Amazon's warehouses and logistics operations, and in particular the operation of Amazon's fulfilment centres. It is in such centres that we are beginning to see some of the most advanced applications of smart-tech working practices, and it is here that we are, perhaps, most likely to catch a glimpse of the opportunities and problems that such systems are generating.

*

The application of smart technology is apparent in all of the key activities associated with the operation of Amazon's numerous fulfilment centres. Amazon fulfilment centres generally operate around six separate stages.[2] First is the receiving of items of supplies. This stage is one that still appears to be highly dependent on humans, as they collect goods from delivery vehicles and scan attached bar codes so that they can become a recognisable part of the smart warehouse. The next stage involves stowing goods in allocated storage spaces. Again, while guided by smart technology, especially in relation to identifying where items should be placed, this is a process that remains heavily dependent on human labour. In certain fulfilment centres robots are now used to move mobile stowing units to human workers (Amazon actually call their workers Associates). This process appears to reduce the need for certain forms of human labour (namely, walking to and from static shelves) but increases the amount of stowing that Associates can complete. In a study of the application of robotic technology in an Amazon fulfilment centre in Germany, it was recognised that related developments could improve safety in the workplace (Fuchs et al., 2021). The use of robot shelf movers does, however, also facilitate something called random stowing, whereby the same objects are not all stored together but in a mobile shelf location that we assume is logistically easiest to reach the human stowers with. This process results in the most popular items being randomly stored throughout the fulfilment centre and makes their retrieval more flexible and efficient. The use of robots to move storage units and randomised storage techniques are interesting from a smart-tech perspective because they reveal how digital technology can optimise a workplace process. But it also has the effect of disconnecting the worker/associate from an understanding of the processes they are involved in. Their skills are primarily reduced to that of a flexible machine, adaptively stowing items. Their cognitive skills to, perhaps, remember where an item is located or to make suggestions about how to improve workplace systems are thus greatly diminished. How can you possibly be expected to creatively contribute to a workplace whose underlying operational logics you cannot comprehend?

The next stage in the fulfilment process involves the *picking* of ordered products. Pickers are the Associates with responsibility for retrieving the items from the smart warehousing system (in the robotic fulfilment centres, the relevant shelves come to the pickers). Interestingly, pickers are not responsible for collecting all of the items in a single customer order. A single order is thus normally collected by several pickers who are selected on the basis of the optimised alignment of robotic shelves and Associates. While logistically optimal, the dissociation of picker from individual customer orders separates Amazon

---

[2]   This account of how an Amazon fulfilment centre operates is derived from this short film that Amazon produced to mirror the physical tours it offers around its facilities: https://youtu.be/UAKPoAn2cB0.

Associates from identifiable people, reducing their role to being a physical conduit between two parts of a smart logistics system.

After picking, orders then proceed to *packing*, where Associates box and tape orders ready for dispatch. Smart technology also infuses the pack area, where orders are automatically allocated a box size, and even the packing tape is automatically cut for size optimisation. By the time a package reaches SLAM (Scan, Label, Apply, Manifest), the human input into the smart warehouse all but disappears, as conveyor belts and automatic labelling systems prepare boxes for dispatch. At the *shipping* stage, humans do re-emerge to load up the lorries that whisk the deliveries away from the fulfilment centre to the army of drivers who bring packages to our homes.

There is one further group of Amazon Associates who support operations in fulfilment centres. These are the Associates and engineers who sit in the *flow room* and observe the operations of the whole system. The flow room is an interesting term, depicting as it does the smooth movement of a logistical process, but also (perhaps accidentally) conjuring up the psychological processes associated with the unthinking execution of complex tasks. Those in the flow room monitor the performance of both the human and mechanical components of the smart warehouse to ensure optimal outputs from this cyborg system.

It is not difficult to depict the digital Taylorism of Amazon fulfilment centres in dystopian terms (see Fuchs et al., 2021). It is important, however, to acknowledge the benefits that are associated with Amazon's smart warehouses. First, it is clear that Amazon is routinely able to deliver goods to customers with unprecedented speed and at relatively low costs. The benefits of this speed of delivery are, in part, a product of the smart operations of fulfilment centres (although perhaps more a result of the hard work of humans). Second, one can only assume that the logistical efficiencies associated with the operations of fulfilment centres are likely to reduce the amount of resources that such facilities need and thus decrease the stress placed on the environment. But, and this is a significant but, it is clear that working in Amazon fulfilment centres has significant human costs: in other words, it is far from fulfilling. In its 2020 report 'Challenging Amazon', Britain's Trades Union Congress (TUC) outline a broad range of problems encountered by fulfilment centre workers (Pound, 2020). According to the TUC, in its drive to put the needs of customers first, Amazon has enacted a series of problematic employment practices (Pound, 2020: 11). The TUC outline gruelling shifts involving ten-hour days and 55-hour weeks (at peak times), comparatively high rates of workplace accidents, and the mistreatment of pregnant workers (Pound, 2020: 12–13).

A major problem that Amazon Associates report relates to expected levels of productivity. The TUC claim that in the UK, Amazon pickers are expected to pick approximately 300 items per hour. Smart monitoring of picker performance can routinely result in disciplinary action and sackings (Pound, 2020: 12). In this context, it is interesting that when surrounded by smart machines, workers' lives are not necessarily made easier. Instead, the increased speed associated with smart logistics and robotics can actually mean that workers are no longer able to set their own pace (O'Connor, 2021). According to O'Connor (2021), the smart workplace is characterised by a dehumunisation and intensification of work within which 'humans are being crunched into a robot system working at a robot pace'.

Related to the punitive actions taken against unproductive Associates are the broader questions concerning surveillance in the workplace. For some time now, prominent scholars such as Alex Pentland have speculated about and explored the use of digital devices (such as mobile phones and wearable badges) as a way of better understanding complex social systems such as workplaces (see Eagle and Pentland, 2006). According to Pentland, the Bluetooth capabilities of smart devices open up new epistemological opportunities for studying social proximity in the workplace (see Olguín et al., 2009). By understanding who comes into contact with whom in the workplace, Pentland (2014) claims it may become possible to determine the forms of routines and relations which foster productivity and those that do not. Ultimately, Pentland (2014) argues that the digital monitoring of social life in any organisation could uncover the social physics of institutions and the associated laws of successful business operations. Although the digital monitoring systems and flow centres used by Amazon have enabled the monitoring of Associates' performance, not even Amazon has been able to realise the vision of total surveillance and social physics envisaged by Pentland. However, COVID-19 and its associated impacts on Amazon fulfilment centres have generated new pressures and opportunities for employee monitoring. To try to prevent the spread of the novel coronavirus in its fulfilment centres, Amazon introduced new smart-tech systems. Amazon has, for example, been using wristband trackers to try to achieve effective social distancing of employees (the wristbands buzz when proximity limits are violated; Pound, 2020). The company has also developed a Distance Assistance technology that provides employees with a visual representation of their social distance with colleagues and whether their distancing circles are being transgressed. This real-time smart tech utilises depth sensors and machine learning to ensure that it can differentiate between people and the background environment. While Amazon's wristbands have not been mandatory for workers, and the Distance Assistant has been designed to guide, not strictly control, associate interaction (BBC, 2020), it is not difficult to imagine how such technologies could be used in more punitive ways. Outside of the COVID-19 context, they could certainly contribute to

Amazon's attempts to further improve the efficiency of its fulfilment centres and put additional corporeal pressures on its workers' activities.

There are different ways in which we could interpret the emerging forms of digital surveillance employed by Amazon. This surveillance could reflect the use of smart technology purely as a method of protecting Associates from the spread of COVID-19 (this would, of course, be in Amazon's economic interests as well). It could reflect the use of smart technology as cover for otherwise lacking health and safety measures and effective leave policies within the corporation (see BBC, 2020). Either way, it seems likely that the introduction of such personalised surveillance tech could be more common in the workplace following its initial trialling during the pandemic. As greater personalised surveillance enters the smart workplace, it is important to note Amazon's expansion into the realm of facial recognition technology (Pound, 2020). This could have broader civil rights implications that go well beyond those of workplace productivity. There is an argument that many of the forms of digital monitoring that are being deployed by Amazon resemble the systems of data surveillance we voluntarily submit to within the domestic sphere. But in the domestic sphere there would appear to be much more choice in terms of the surveillance that we are willing to tolerate. It is also clear that in the domestic sphere dataveillance is not likely to result in the termination of employment.

*

The case of Amazon's fulfilment centres speaks to a broader issue concerning the different operational forms of smart tech. As we have previously observed in the domestic sphere, smart tech can learn from users (and broader user populations) to provide optimised, predictive responses for those users. Of course, this kind of learning is always open to commercial exploitation. But the smart systems we see in Amazon's fulfilment centres appear to follow a different operational logic. While it may be able to learn from worker input, in the long term this approach appears to involve the imposition of optimising logics onto workers. This optimising logic is not about making Associates' working lives better (though that could be a consequence). Rather, it is about the optimising of cost and time in the service of profit maximisation and customer service. In this context the smart learning evident in fulfilment centres is not about machines and algorithms learning about humans to make human life more convenient. It is about machines and algorithms learning about the interactions between humans and machines and attempting to optimise the commercial benefits of these interactions.

From Dark Factories to the Optimisation of Dignity

There is an emerging body of work that explores the ethical dimensions of smart-tech workplaces. In the aforementioned World Economic Forum's 'The Future of Jobs Report 2020', it is claimed that we are currently in a narrow window of time when we have the opportunity to shape smart-tech work in humane directions:

> As the frontier between the work tasks performed by humans and those performed by machines and algorithms shifts, we have a short window of opportunity to ensure that these transformations lead to a new age of good work, good jobs and improved quality of life for all. In the midst of the pandemic recession, this window is closing fast. Businesses, governments and workers must plan to work together to implement a new vision for the global workforce. (2020a: 4)

But the question still remains as to what an ethical application of smart technology in the workplace would actually look like.

We can, of course, hope that where smart technology does displace human labour it can be targeted on those areas of employment that are dull, dangerous, and/or dirty (Darling, 2021). In her discussion of the deployment of robotic technology, Darling makes the following observation:

> In most cases, the outcomes are better when robots work with people. In well-defined spaces, like rows of crops, robots are able to take on more of the work. In other areas, like delivering a hamburger in San Francisco, they need a ton of human help to deal with the unexpected. Rather than viewing these limitations as a tricky phase on the way to human replacement, we should stop and ask why are we trying to recreate human skills at all? Why are we trying to replicate something we already have? (2021: 15)

As we have argued elsewhere in this volume, the adoption of smart technology does not necessarily have to undermine human autonomy and dignity. When sensibly deployed, smart tech can produce outcomes that are both socially and economically beneficial. As Darling again states,

> The more fruitful path is to explore what else we can come up with. Where robots truly shine isn't in replacing the college student who delivers pizza. They're most powerful when their form and function helps us to do things we can't do very well ourselves, or even at all. (2021: 15)

If humans are to form workplace partnerships with smart machines and programmes that are socially progressive, as well economically desirable, careful attention must be given to how the smart workplace is constructed. In a survey of UK workers carried out by the TUC, 74 per cent felt that the use of smart tech could enable them to gain greater control over their working lives (perhaps through the saving of time, better understandings of the operation of the workplace, or the ability to focus on more creative and rewarding tasks; O'Grady, 2021). But in the same survey workers expressed concern over the misapplication of new technology, with only 31 per cent being actively consulted about the introduction of novel technologies in their workplaces (O'Grady, 2021). As is often the case, the dangers of introducing smart technology into workplaces are not necessarily a product of the technology, but the wider social context associated with its introduction. In countries like the UK and US, the introduction of smart tech into workplaces corresponds with a time of weakened workers' rights, pay, and conditions, and the declining influence of workers' unions (O'Grady, 2021).[3] It is much more likely in such circumstances that smart technology will become a socially disempowering force in the workplace, a force that further erodes the ability of humans to control key aspects of their working lives. It is particularly troubling that Amazon is not only not consulting Associates about the introduction of smart technology, but also using smart tech to undermine the impacts of industrial action (Fuchs et al., 2021). According to Fuchs et al. (2021), Amazon is using the algorithmic insights it has gained from cold weather days—when certain fulfilment centres may see their productivity drop perhaps due to icy conditions, or workers not being able to get to fulfilment centres—to anticipate and mitigate against the impacts of striking workers.

A broader sense of what is at stake within the transition to the smart-tech workplace is provided in Pasquale's (2020) analysis of the relationship between human expertise and AI. Echoing the work of Darling, Pasquale argues that the preservation of fulfilling vocations is central to human dignity and a related sense of purpose. Focusing specifically on the question of expertise, however, Pasquale lucidly argues that the preservation of professionalism is central to broader forms of human empowerment. For Pasquale, professionalism is not merely about being good at your job, but also being able to meaningfully contribute to discussions concerning how that job is best delivered. For Pasquale, then, professionalism is not just a set of technical skills; it represents a form of social empowerment that enables people to participate in determining the conditions associated with working life (2020: 4). Even if human expertise is economically suboptimal, we may choose to protect it as a route to preserving human empowerment. But there may well be other reasons to protect and enhance human professionalism in a smart-tech age. Workers report that being

---

[3]   Fuchs et al.'s (2021) study of Amazon fulfilment centre staff in Germany reveals that they were rarely consulted about the application of new technology in the workplace.

managed by an algorithm is lonely and pressurised and is likely to thus have economic disadvantages in the long term (O'Grady, 2021). Additionally, Pasquale reflects on the role played by humans in the creative industries (2020: 5). Perhaps algorithms may ultimately be able to pick the film scripts that are most likely to be commercially successful, but it is less likely that smart tech will be able to select the scripts that are likely to have the greatest cultural value in the long term (Pasquale, 2020: 5).

It seems unlikely that workerless dark factories will be the most effective way of organising economic activity or delivering broader social needs in the future (Ball, 2021). But there is much work to be done to ensure that the smart-tech workplaces of the future are actually places that humans want to live out their working lives.

THE GOVERNMENTAL AMBITIONS OF SMART TECH

This chapter has so far focused on the emerging use of smart technology within the internal operations of corporations and the associated implications for human labour and expertise. The second half of this chapter considers the ways in which smart technology is opening up new opportunities for corporations to intervene within the public realm. This section is thus a prelude to the broader discussion of the governmental use of smart technology we develop in the next chapter. Here we shift our focus from a concern with the connections between smart technology and the governance of employees to consider how smart tech appears to be enabling corporations to assume broader governmental roles within social life. As we will see, smart technologies have enabled corporations to generate new opportunities for acts of corporate governance. Specifically, the data-gathering capacities of smart-tech companies facilitate ways of seeing the world, which reflect and, in some instances, surpass the oversights associated with nation states (see Scott, 1999). Through the application of digital nudging and psychological influence, smart-tech corporations have also started to rival the monopoly of behavioural control associated with state bureaucracies (see Chapter 4, this volume). In addition to generating new capacities to govern, however, it is also apparent that the social problems created by Big-Tech corporations have generated new things that need governing (for example, the moderation of content on social media platforms). Smart-tech corporations are thus simultaneously creating new governmental problems and developing new capacities to govern them. In what follows, we consider how these developments inform the broader social and political realignments that are associated with the smart-tech society.

The Novel 20th-Century Corporate Narrative

First, we provide some context and a slight digression into early corporate purpose and corporate law. The 20th-century norms concerning the role of the corporation within society are something of an historical aberration. In their early history, corporations operated very much in the shadow of the state. It was states, through acts of charter, that gave authority to corporations to govern various spaces of life on the individual and societal levels. In this context, corporations were originally formed to 'play a fundamental role in providing public goods and exercise powers customarily associated with formal state institutions' (Barkan, 2013: 1). The purpose of the corporation has, however, changed significantly over time. The modern iteration of the corporation has its origins in US corporate law. This legislation changed the face of the corporation granting and establishing the powers to hold property, limited liability but most importantly the legality of holding companies (which meant that companies could own stock in other companies). This changed the face of the corporation and effectively ended its historical relationship with the state. The revocation of the grants of charter meant companies would no longer be tools of the state that were meant to 'directly reinforce the public welfare' (Barkan, 2013: 57). The dominant narrative of the 20th century thus became that the corporation was to provide for the needs for a small band of shareholders.

Towards the end of the 20th century a shift began to emerge in the operating logics of many corporations. In an interview we conducted, the Chief Knowledge Officer of a leading consultancy firm described this transition in the following way:

> [we are in] a period of time in which people's expectations about brands have shifted again. They shifted once from product to person and they're shifting now from person to public. They're not leaving the product or person behind, they are just adding, so it's kind of a layering on top ... but now there's a greater expectation that brands also have to do something for the public. Brands have to deliver something that contributes to social wellbeing or society at large ... the underlying equation hasn't changed, we're still trying to do what it is that consumers want, you're still trying to deliver what consumers expect of your brand. You're still trying to meet consumer needs. It's just consumers had new needs, consumers have new expectations ... There's a change in people's perceptions of government as the traditional stronghold of doing something about society. (Chief Knowledge Officer interview, 2020)

The consumer demand for Corporate Social Responsibility outlined here can be seen as a form of compensation for the socio-environmental harms created by corporations. In this context, they predate the onset of the smart-tech society. As we will see, however, smart technology has generated new harms that require corporate governance. We also claim that smart technology has enabled forms of corporate engagement in governmental affairs that go far beyond the realms of Corporate Social Responsibility.

The emergence of an expanded government role for corporations in the smart-tech society is in part a product of the neoliberal retrenchment of the state. Returning to our conversation with our industry insider,

> Governments have abandoned a lot of the things that they have done historically … people are looking to brands. People not only want a great product, not only want something that makes them a better person but something that also contributes to a better society and/or better public. That is changing what it is that brands have to do in order to find themselves in the consumers' considerations. (Chief Knowledge Officer interview, 2020)

The emergence of a smart-tech society is thus synonymous with a point in time when the capacity of many Western states to govern has been fiscally eroded. It has also corresponded with an emerging desire of corporations in general to engage in public matters. Finally, the smart-tech society is also characterised by a time when Big-Tech corporations have unprecedented reach and often a monopoly of control over shared digital spaces. It is in these contexts that we find smart-tech firms, such as Uber (public mobility), Twitter/Facebook (speech and expression), and Amazon (logistics and supply), becoming significant players in not only commercial but also public life. The smart-tech society is once again redefining the evolving historical relations between the corporation and the state. In the smart-tech world, though the corporation is no longer separate from government or subservient to the state—it is increasingly occupying the role of government!

Smart-Tech Government: Between Trespass and Trust

The smart-tech corporate occupation of the realms of government can be interpreted in three broad ways. First, it can be seen as one of a series of smart-tech trespasses. According to Zuboff (2019), the right to trespass is one of the defining characteristics of the Big-Tech sector. Zuboff (2019) describes the notion of trespass as enabling surveillance capitalists to expand their data reach rapidly without having to gain legal authority or consent. The idea of the right to trespass can be applied to surveillance capitalist occupation of governmental arenas. Many Big-Tech firms are now assuming governmental roles in the field of public health in response to the COVID-19 pandemic (see, for example, Facebook's 'Data for Good' COVID-19 programme). These activities are often undertaken without direct governmental oversight or consent by the public who are being subject to corporate forms of government.

But there is more to Big-Tech acts of government than mere trespass. While there is clearly public concern surrounding the level of smart-tech intervention into individuals' daily lives, there is an evident shift in public sentiment globally about the private sector. Edelman, a global communications firm, has been studying trust for the last 20 years in the four major social institutions: government, business, media, and nongovernmental organisations. In 2020 there was a significant change in precedent. Business became the only trusted institution among the four. Corporations were also the only institution seen as both competent and ethical. With a decline in trust in government institutions and increasing trust in the private sector, it is possible to argue that there is an implicit legitimacy to the smart-tech corporate occupation of the realms of government. The Edelman report (2021) indicates that over 65 per cent of people agree with the following statements: CEOs should step in when governments do not fix societal problems; CEOs should take the lead on change rather than waiting for government to impose changes on them; and CEOs should hold themselves accountable to the public and not just the board of directors or stockholders. These sentiments seem to reflect an increasing willingness among the population to be governed by corporations. What is clear is that smart-tech corporations have the reach and technical capacity to take advantage of these changes in public opinion. It is also apparent that the smart-tech business models will be able to effectively maximise potential gains that will accrue from the assumption of governmental roles.

It is perhaps ironic that a third interpretation of the emerging governmental role of smart-tech corporations suggests that Big-Tech is increasingly being called upon to address public problems of its own creation. The most obvious example of this situation is evident in the case of social media platforms. Over time, the growth and network effects of social media such as Twitter and Facebook mean that they are crucial sites for public speech and knowledge sharing. While offering the opportunity for a global reach of individuals' speech, such platforms have also generated ample opportunity for the abuse of acts of free speech. Grimmelmann summarises these abuses in four categories:

> congestion, which makes it harder for any information to get through ... cacophony, which makes it harder for participants to find what they want ... Both congestion and cacophony are problems of prioritization: bad content crowds out good ... Next, there is abuse, in which the community generates negative-value content— information 'bads' ... Finally, there is manipulation, in which ideologically motivated participants try to skew the information available through the community. (2015: 53–4)

The scale of social media platforms means that effective content moderation by human actors is impossible. To put this in perspective, the volume of photos uploaded to Facebook on a daily basis is in the region of 350 million per day (Smith, 2019). If we take this number and (optimistically) suppose that a moderator would be able to look at and decide if a photo is appropriate every second of an eight-hour work day, that would mean that Facebook would need 12 153 moderators to check for abusive or misleading content in just one category of content hosted on the platform. The proposed solution to the problem of content moderation associated with smart-tech platforms is thus smart tech. In this case, it is the development of automated filtering to remove content abuses or to send the content to human moderators if the content is not immediately identifiable as harmful. The regulation of free speech is something that has historically been associated with governmental oversight, legislation, and policing. In the smart-tech society, though, smart tech generates a heightened need for governmental forms of regulation of speech acts, but it also appears to offer the only viable technique of enacting effective government. Similar situations emerge when rideshare platforms such as Uber simultaneously generate passenger safety problems for taxi users and smart-tech solutions to solve those same problems.

Smart-Tech Corporate Government in Action

Having established the historical context and frameworks of explanation for smart-tech corporations' acts of government, we now consider more specific practical manifestations of this phenomenon. An interview we conducted with a business leader reveals the nature of what we are exploring:

> Think of the government as old-school London: narrow roads, old sewage systems, which still work. There's nothing wrong with them, they still work. They're still functioning. But then you look at the likes of new modern cities, the likes of Milton Keynes … it's all built in terms of modern infrastructure. That's how I see the tech platforms, in terms of an analogy. The efficiencies are huge … The thing is, the tech brands have the data, they know how to use that data, to a certain degree. But the scary part is the scale at which they can grow, due to AI and other technology, what [companies] … already do, is the scary part to me. And the fact that governments are not thinking in an agile way … Think in terms of government, the new age of businesses working together. So, you look at how brands are becoming less hierarchical and more agile in their approach to business and what they're trying to be; it's more about the human, the individual, it's a lot more nimble, in terms of its approach. (Business insider interview, 2020)

This analogy of government institutions and the capacities of corporations in infrastructure terms provides a concise and intuitive explanation of the nature of the developments we are interested in here. The issue that is revealed here is not just the smart-tech corporation's occupation of the realms of government, but the transformation of how governing is done. To develop some analytical perspective on these processes, we now consider some specific examples of smart-tech corporate government in practice.

The first example is one that readers may be familiar with—that of dynamic and personalised pricing. Dynamic pricing refers to variable pricing on services or products that reactively change prices due to a specific set of conditions. An example that many people will have encountered is used by Uber. Prices for Uber change on demand; this is called surge pricing. The higher the demand, the higher the cost for hailing an Uber. Price here is algorithmically determined by the requests that are received, and prices change accordingly. Uber (n.d.) states, 'because rates are updated based on the demand in real time, surge can change quickly. Surge pricing is also specific to different areas in a city, so some neighbourhoods may have surge pricing at the same time that other neighbourhoods do not'. This system is not unique to Uber; dockless bike-sharing companies like Mobike in China use similar smart technologies to provide preferential rates to ensure they manage the supply of bikes across the city. Demand for bikes at different times of day means that bikes need to be redistributed back to places with higher demand; this can be done through preferential pricing for rides to locations of demand. These forms of smart-tech variable pricing have little to do with corporate acts of government. In other contexts, however, their application has taken a more governmental form. Some bike-share companies, for example, have been trialling personalised behavioural pricing models for deterrence of antisocial behaviour. A CEO of a retail market insights firm observed to us how,

> If you left the bike on the side of the road in a bad position or rode it on the sidewalk or in a place where you shouldn't be riding it, or went down the left side of the road when you should have been on the right side of the road, or vice versa, it would detect that you did that. Then the next time you would rent a bike, instead of paying 50 cents per mile, you'd pay a dollar per mile. (Retail Market Insights CEO interview, 2020)

There is a clear economic incentive for bike-share corporations to use smart pricing to support prosocial cycling behaviour. It is, however, also clear that such actions reflect novel forms of corporate intervention into the governing of user conduct in public space—an action space that has historically been reserved for the state.

In some cases the use of smart-tech systems by corporations does not lead them directly into realms of government, but it does blur the boundaries of responsibility between corporations and various branches of the state. A case in point is provided by the fast-food giant McDonald's. McDonald's have been trialling the use of a smart-tech facial recognition in its drive-throughs as a way of improving the efficiency of food delivery. The CEO of McDonald's in Russia described the innovation in this way:

> We [McDonald's] have facial recognition and we basically use it at the drive-through because the drive-through is our number-one product. It's the number-one way that people get a burger or whatever they order, and what we use it for is just to speed up the order process. So, for example, if a car drives up to the McDonald's and there's 30 cars in front of it, we can identify the licence plate and then the people in the car to see what are their patterns. And if there's a pattern that says whenever they come here they definitely always order a black coffee or an espresso, we can make sure that the coffee machine is topped up and ready to go so that there's no delays in fulfilling the orders. It sounds kind of, like, very good use of personal data, right? (Retail Market Insights CEO interview, 2020)

On the surface, this use of smart surveillance may seem helpful and convenient. But this technological process has implications beyond the efficient delivery of fast food:

> The problem is you're violating potentially two different sets of data rules [and potentially rights]. The first, now that you've got their licence plate, the car, you can tell if somebody is a criminal. So, somebody who shouldn't be driving the car is driving the car. You potentially could be flagging that somebody is driving a stolen car, which is a criminal offence. (Retail Market Insights CEO interview, 2020)

What then is the obligation of the retailer who has identified a criminal or criminal behaviour? Should it be reported? Is the restaurant obligated to pass this information on? In this instance, it is possible to think of McDonald's drive-throughs as becoming branches of the state's law enforcement systems. Of course, such developments reflect a much larger set of connections between Big-Tech and state surveillance. But in this instance, through the application of smart tech in the real world, corporations have the capacity, and potentially the responsibility, to support public policing (whether they want to or not).

Emerging patterns of smart-tech development suggest that the boundary between corporations and policing is likely to become ever more blurred in the future. The development of 5G technology will increase data upload and download capabilities 100 times (from 100–300 Mbps to 10–30 Gbps!; iSelect, n.d.). There are three main forms of digital data transfer that operate in our everyday lives. The first is continuous data uploads such as those associated with traffic cameras, whereby if you run a red light, your data are automatically transferred and you get a ticket. This continuous data, however, only makes up approximately 0.3 per cent of data on devices. Second are those data stores that can be set to transfer on a schedule, such as backing up your messages to the cloud on a weekly basis. The last category is data that only gets tapped into when diagnostics are needed, such as your car's onboard computer that measures your tyre pressure or the number of times you turn on your windshield wipers. These data are only used for diagnostics when your car breaks down or there is a problem. These last two categories make up the other 99.7 per cent of data stored locally on devices (Retail Market Insights CEO interview, 2020). With 5G technology there will be a vast rebalancing in the percentages of data that can be cheaply and efficiently accessed in real time:

> If you buy a car five years in the future, there's a very good chance that it's going to be communicating continuously with the police because the roads will have smart technology installed on them. The car will have smart technology ... So you'll have continuous data, real live time. All that [data is] both on the local device and on the non-local device, meaning the road itself or the traffic camera itself. (Retail Market Insights CEO interview, 2020)

It is highly likely that in the future, your smart car could be sharing diagnostic data with police forces to ensure that your car is safe and roadworthy, or liaising directly with car owners to ensure they are not in breach of traffic regulations. In this context, even if car manufacturers and their smart-tech systems are not involved directly in governing driver behaviour, they will certainly be part of the network of digital systems that do. This is a novel, if not unprecedented, role of corporations, which has implications for how we understand the separation of powers and influence between states, corporations, and civil society. In these contexts, it is not just conceivable but likely that the smart corporations will end up becoming a central actor in the provision of health, public order, welfare, and prosperity for vast swathes of the population.

CONCLUSION

In this chapter we have outlined some of the ways in which smart technology influences the operations and roles of corporations. In the first half of the chapter we explored how smart-tech systems are reshaping working life within corporations. The full implications of smart tech for the job market are as yet unclear, with certain sectors likely to need less human input, but others likely to see more. The operation of Amazon's fulfilment centres does, however, reveal that even within one corporation the nature and impacts of the smart-tech–human interface will vary greatly according to precisely where you work in the organisation. The example of Amazon also revealed one of the inherent dangers of the smart-tech workplace. Rather than making working life less dangerous and demanding, and potentially more fulfilling, there is a clear risk that the smart-tech workplace could result in the expectation that humans become increasingly machine-like in their working practices and lives. It also appears that the presence of smart technology in the workplace will present a challenge to human expertise and dignity. The use of smart technology to produce efficiencies in working life is likely to result in a diminished capacity of humans to be able to comprehend the full nature of the working processes they are a part of. This will unquestionably be a threat to human claims to professionalism and expertise. In turn, this could result in less human input into the design and orchestration of workplace systems. This threat, perhaps more so than the loss of work, is arguably the most significant issue associated with the smart-tech workplace. It raises the question of how we can collectively ensure that workplaces are designed to preserve (and hopefully enhance) human respect and fulfilment as well as maintain profit margins. The socio-technological window to secure human input into the smart-tech workplace appears to be closing rapidly.

In the second half of this chapter we considered the emerging governmental role of corporations. It appears that the rise of smart tech, and associated forms of digital data surveillance, are enabling a blurring of the line between corporations and the actions of states. The governmental activities of corporations appear to be the outcome of both design and chance. With the emergence of historically unprecedented sources of behavioural data and influence, corporations are uniquely placed to be able to guide public behaviour. At the same time, the social problems generated by smart tech necessitate new forms of governance, which often only Big-Tech corporations are able to deliver. Sometimes, however, it appears that the governmental role of smart-tech corporations is the product of unplanned alignments between digital data gathering and the enforcement of the law. In some ways the enhanced involvement of corporations—long castigated for their selfish antisocial gestalt—in public life should be celebrated. But vigilance is surely required here. If corporations start to look and act a lot like governments but have none of the attendant democratic checks and balances on their activities, should this not be a concern? As with the smart-tech workplace, it is thus critical that powerful lines of human input and accountability are hard-wired into the digital worlds we are building.

7.   Smart-tech states

INTRODUCTION

On the evening of 11 March 2020, the UK prime minister convened a curious meeting. In response to the emerging COVID-19 pandemic, Boris Johnson gathered experts to try to help the government tackle the crisis. What was peculiar about this meeting was that it did not involve the usual collection of epidemiologists, virologists, and public health experts. Instead, the meeting drew together representatives from the smart-tech industry. Google's DeepMind artificial intelligence research unit had a representative there, as did the shadowy data analytics company Palantir. Apple, Amazon, Microsoft, and Uber were among the 30 or so other Big-Tech companies who attended. This gathering of the great and the good in Big-Tech was described by the press as a 'digital Dunkirk': a call from the British state for urgent help in a time of crisis (Waterson, 2020).

The commercial benefits of the COVID-19 pandemic to Big-Tech have already been outlined (Galloway, 2021). However, the digital Dunkirk meeting reflects something more than just short-term commercialism. It is also distinct from the expanded governmental role of smart-tech corporations in our everyday lives outlined in the previous chapter. It embodies a formal fusion between the apparatuses of public government and smart technology. This fusion predates the COVID-19 pandemic but has clearly been accelerating within the states of exception associated with the governmental responses to this crisis (Amoore, 2020; Isin and Ruppert, 2020). It appears that in the context of an ever more complex set of wicked problems that are being faced by governments around the world (including climate change, public health, cybercrime, and terrorism) that state authorities are interested in the analytical and behavioural power promised by smart technology. This fusion between government and smart tech is also being driven by the desire of Big-Tech companies to expand their digital surveillance capacities and analytical prowess by piggybacking on the institutional and infrastructural reach of states. In these contexts, smart technology is increasingly being used by government authorities to identify socio-economic problems, design and deliver policy solutions, and monitor the efficacy of related government interventions (Lago, 2021). Smart tech is already being used to govern at a range of policy scales, including local outbreaks of public unrest and riots, urban transport systems, and national immigration policies. These processes reflect novel scales of digital monitoring, analysis, and feedback. They are the manifestations of the smart-tech society at a collective level.

*

So far in this volume we have mainly considered the use of smart technology at an individual level and in the context of the private spaces of the home and corporation. This chapter explores the emerging uses of smart technology within public life. We draw particular attention in this context to the applications of smart tech within the arena of government. When we use the term government, we are not only referring to the institutions of national and local governments (including nation states and city authorities). We are also drawing attention to the processes through which social order is pursued. As the broad goal of all systems of government, the pursuit of social order can be discerned in varied contexts ranging from the short-term policing of a protest to the longer-term management of a pandemic. The pursuit of social order does, of course, take different forms, and has very distinctive purposes according to the context within which it is pursued. In liberal democracies the pursuit of social order is generally associated with the securing of economic prosperity alongside the maintenance of personal freedom. In more authoritarian contexts social order is synonymous with consolidating the power of a single leader and/or party, while the preservation of freedom is of limited concern.

As the case of the UK's digital Dunkirk meeting demonstrates, it is clear that the emergence of COVID-19 has served to justify and accelerate the use of smart technology within systems of government around the world. In this chapter we consider the social and political implications of a deepening fusion of digital technology and government. At a social level, we explore what it is like to be governed within the learning loops of smart technology and the associated experience of interacting with digital systems as opposed to human representatives of the state. In a political context, we explore the constitutional implications of smart-tech government and what it means for questions of accountability, transparency, and political participation.

This chapter commences by considering how we might begin to think and theorise about the notion of smart government. In this section we argue that while the nature of the smart-tech state may not be as historically novel as you might think, the emerging fusion of government institutions and digital technology does reflect an important shift in the nature of statehood. The two sections that follow then consider the practical application and experiences of smart-tech government at two scales. First, attention is given to attempts to use smart technology within the national welfare programme of the UK. Second, the analysis explores the use of smart technology at an urban scale through the case of Singapore.

SMART GOVERNING—SOME HISTORICAL AND PRACTICAL CONTEXT

Smart Tech and the History of Modern Government

To discuss questions of government and public order is to enter a rarefied arena of philosophical debate that encompasses the work of Plato, Hobbes, and Rousseau (among many others). The relationship between government and social order are central concerns within philosophy. They are central because they speak to fundamental questions concerning the nature of the human condition (in particular, our moral tendencies towards good and evil, and conviviality and self-interest), and how best to secure freedom when the expression of one person's liberty can undermine the freedom of another.[1] While we cannot delve too deeply into this complex history of ideas, to explore questions of government and social order in the smart-tech society requires us to ask at least two fundamental questions. First, what is the purpose of government and associated systems of social order? Second, how are government and social order to be achieved?

An instructive starting point when exploring these questions is the work of the French philosopher Michel Foucault. Foucault was interested primarily in the history of ideas associated with the human sciences (particularly those connected with psychiatry, criminology, and sexual conduct). A significant part of his later studies of the genealogies of knowledge focused on the sciences of government (Foucault, 2008). Foucault's work is of significance to our discussion because he identified distinctive purposes and practices within the history of government. It is also significant because his analysis reveals the complex, changing, and often overlapping goals and techniques of government. According to Foucault, acts of government may focus on questions of sovereignty, as a monarch or leader uses their power to defend their territory and assets. Government can, however, take a more holistic and interventionist form as states seek to order the actions of citizens through mechanisms of disciplinary control, which often carry with them the threats of violence and death. Foucault's primary interest, however, was with a set of governmental practices that have come to predominate (particularly in liberal democracies) since the 17th and 18th centuries. These are the systems of government we are

---

[1]   For an excellent overview of these debates and a history of related schools of philosophical thought, see de Dijn (2020).

most familiar with but are often oblivious to. They are systems of government that are associated with pastoral care and the careful monitoring of national populations. Foucault uses the rather confusing neologism 'governmentality' to describe these practices. The 'mentality of government', which Foucault argues has emerged in liberal democracies, is one that is concerned with more than the narrow interests of a sovereign and generally shuns the disciplinary tactics of a police state. It is an approach to government that gains power and legitimacy through support for life, rather than through the threat of violent death.

The systems of governmentality identified by Foucault reflect a response to scientific and political developments in the 17th and 18th centuries. The scientific revolution delegitimised the religious authority upon which monarchical sovereignty was often based. This so-called Age of Enlightenment also emphasised the importance of human liberty and dignity, which brought into question the arbitrary exercise of disciplinary forms of power. In these contexts, governmentality sought to redefine the purpose of the state by positioning it as an instrument targeted at supporting human wellbeing and liberty. Despite such lofty ideals, Foucault argues that governmentality is a pastoral form of government only to the extent that associated systems of collective care are conducive to the needs of the market economy. The purpose of governmentality is thus to attain social order through the provision of human welfare that is conducive to economic growth and prosperity. But how is this to be achieved? According to Foucault, Age of Enlightenment liberal forms of governmentality are synonymous with very specific techniques of governing. Drawing inspiration from the scientific revolution, governmentality is first and foremost predicated on the deepening of the connections between governmental power and knowledge. Governmentality thus relies on the development of various data-gathering techniques, including those that collect economic records, health registers, censuses, and more generalised forms of surveillance. Scott (1999) has argued that in this context modern governments have become synonymous with certain ways of *Seeing Like a State* (the title of his 1999 book). These ways of seeing have relied upon standardisations in the ways in which all manner of things are measured. Governmentality was thus enabled by, and a facilitator of, the birth of the science of stat(e)istics. Beyond the construction of new regimes of knowledge, governmentality also resulted in the emergence of novel practices of state intervention targeted at human wellbeing. Public health care, improved urban sanitation systems, and universal education are all facets of governmentality.

The relationship between smart technology and Foucauldian theories of governmental power is contested and undoubtedly complex. We dwell briefly on these connections and contestations here because they begin to reveal some of the significant political implications of smart government. At perhaps the most rudimentary level, Foucauldian theories of governmentality and smart govern-ment are connected by the emphasis they both place on the relations between power and knowledge. Of course, the knowledge (data is a better word) gath-ering and analysis capacities of the smart-tech society are more extensive and sophisticated then Foucault could ever realistically have anticipated (dying in 1984, Foucault would have been aware of the emergence of computing power, but not of where the still novel notion of personal computing was heading). Despite the obvious connections between theories of governmentality and the practices of smart government, there is still much conjecture concerning the continuities and discontinuities between these realms of government. These conjectures are at the centre of new academic concepts—such as the notion of algorithmic governmentality.

According to Amoore, algorithmic governmentality involves '[l]ogically governing uncertain human actions and behaviours through relations between data points' (2017: 1). For Amoore, prefixing governmentality with an algo-rithmic adjective indicates a novel disjuncture with the analogue systems of governmentality outlined by Foucault. While both systems of governmentality are knowledge oriented, algorithmic governance embodies a shift from the deductive logic of a governing hypothesis, which is tested through available data, to an inductive rationale (Amoore, 2017: 2). Through continuous moni-toring and the accumulation of data correlation points, algorithms enable gov-ernments to respond to emerging trends (which may be fostered or suppressed) and govern in a more open and anticipatory way. In this context, Amoore argues that data are not the end of the governing loop, but rather act as its gen-erative starting point. Algorithmic government is a relatively open and experi-mental form of government that does not govern through pre-existing theories of reality but rather on the basis of emergent socio-economic trends. Perhaps the most novel aspect of algorithmic governmentality is the way in which it appears to transform the problems of knowledge within government. Amoore puts it thus: '[v]olumes of relatively unstructured data were transformed from a governmental problem of what could not be grasped and known into a means of precisely governing the unknown and uncertain' (2017: 2). In this context, algorithmic governmentality challenges the (neoliberal) limitations that have conventionally been placed on government (namely, that it could not know enough, particularly about market transactions, to govern effectively) by trans-forming knowledge abundance from a weakness to a strength of governing.

Cooper (2020) emphasises the continuities that exist between governmental-ity and its algorithmic descendants. Foucault claimed that the model for gov-ernmentality was the pre-modern Judeo-Christian tradition of the pastorate: namely, the shepherd tending to their flock (Foucault, 2008). In this model, the *shepherdic* state is concerned with the welfare of each citizen and the wider population, just as the shepherd tends to each sheep and the wider herd.

According to Cooper (2020), in a peculiar twist of historical fate, the rise of algorithmic systems of government, with their enhanced capacities to monitor each and all, reflects the fulfilment of the pastoral model of government. It is instructive to reflect on the particular forms of continuity which Cooper identifies between the pastorate and algorithmic governmentality because they help to reveal significant aspects of smart government. First, Cooper emphasises the ways in which algorithmic systems of digital government possess the ability to continually monitor and guide (through behavioural interventions) the actions of citizens. These forms of intensive surveillance and guidance are signatures of pastoral power. Significantly, these are aspirations of government that could never be realised in the analogue manifestations of governmentality identified by Foucault. Second, Cooper (in a similar way to Amoore) recognises that algorithms do not only enable governments to respond to emerging social processes, but also enable state authorities to predict and emergently shape the world. According to Cooper, the anticipatory shaping of the world is a hallmark of pastoral care and the key basis through which the shepherd protects their flock. Third, Cooper recognises a dehumanising tendency that connects pastoral models of power and smart-tech governance. The religious foundation of pastoral power indicates the importance of subjects renouncing autonomy to avoid falling into a state of sin. In the algorithmic era, human competence and autonomy are brought into question by behavioural and psychological sciences (see Chapter 4, this volume). If the pastorate sought to supplant human autonomy with the word of God, smart-tech government replaces it with the algorithm.

There is however an apparent discontinuity between algorithmic governance and pastoralism, which Cooper appears to overlook. Cooper argues that the steering of souls is a central aspect of both pastoralism and algorithmic governmentality. It does seem that systems of smart government could be, and sometime are, targeted on moral sensibilities and the beliefs of citizens. According to Zuboff, however, the algorithmic age is characterised by a new species of power: what she terms *instrumentarianism* (Zuboff, 2019: 351–75). Instrumentarian power is not interested in shaping and guiding souls, but in experimental conditioning and behaviour modification. Here we see that unlike pastoral models of power, smart governance does not have to be delivered through up-close systems of moral instruction, guidance, and monitoring. It is a form of government that can be delivered remotely, through reshaped digital choice architectures and hypernudges.

*

It can be argued that smart government (as expressed by notions of algorithmic governmentality) embodies both the fulfilment and novel extension of the systems of governmentality identified by Foucault. Whether we focus on indicators of continuity or discontinuity, it seems clear that while smart government may have the same broad goals as governmentality, its techniques for achieving them are distinctive. Perhaps the earliest thinker to grasp the societal implications of digital technology on governmental power was Gilles Deleuze. A contemporary of Foucault, Deleuze was also a French philosopher. In a short but influential 1990 essay titled *Postscript on the Societies of Control*, Deleuze charted the rise of new modalities of power that were synonymous with machines like computers. As with Foucault, it is clear that Deleuze could not grasp fully the implications of computing technology. But writing six years after Foucault's death, he was able to discern the likely impacts of digital technology on systems of government and social control. At the centre of Deleuze's (1990) analysis is a purported transition from a disciplinary society to societies of control. Drawing on the work of Foucault, Deleuze identifies the 18th and 19th centuries as the era of disciplinary societies. In these societies, governmental control was primarily administered through institutions of enclosure, such as schools, factories, prisons, families, clinics, and barracks. These spaces sought to directly order and shape social conduct through the establishment of norms and learned practices. According to Deleuze, however, while these institutions did not disappear during the 20th century, they were gradually overwritten by another system of collective power.

Deleuze's society of control is characterised by more open systems of government, within which individuals are afforded greater freedom through the presence of intermittent examination and self-governance. The rise of computer technology is central to this more open style of government. At the same time, computer technology enabled individuals to be liberated from traditional enclosures while also producing the data upon which assessment of individual performance could be gathered and relayed at a distance. Deleuze used the following analogy to describe the different modes of operation of disciplinary and control societies: 'Enclosures are *molds*, distinct castings, but controls are *modulation*, like a self-deforming cast that will continuously change from one moment to the other, or like a sieve whose mesh will transmute from point to point' (1990: 4, emphasis in original). According to Deleuze, then, societies of (digital) control enable individuals to be governed while free of historical constraints on movement and personal comportment.

Deleuze's *Postscript on the Societies of Control* certainly seems to be what is coming down the digital pipeline in the 21st century. His analysis is instructive, however, not only because of what it anticipates but also because of what it remains blind to. For Deleuze, the Society of Control's fusion of freedom and power was a product of intermittent examination. The use of intermittent examination (perhaps training or selective surveillance) means that citizens can be trusted to behave appropriately not because they were being constantly monitored and disciplined but because, in anticipation of inspection, they would self-regulate their behaviour. But in the smart-technology era, new governmental opportunities have emerged. The spread of internet-connected digital devices now means that the whole world has become like a surveillance enclosure. The smart-tech society affords continuous opportunities to monitor, condition, and shape behaviour. This has not, of course, resulted in the return of disciplinary social practices (at least not everywhere). As we will explore, the smart-tech society appears to reflect the self-regulating freedoms of Deleuze's Society of Control along with the novel capacity for disciplinary action. In the smart-tech society, intermittent examination gives way to the perpetual experiment (see Chapter 4). As Isin and Ruppert (2020) argue, Deleuze's Society of Control may, in part, actually reflect the continuation of disciplinary power by other means.

*

The use of smart technology within the governance of the COVID-19 pandemic has led to new theoretical speculation concerning the implications of digital technology on emerging forms of governmental power.[2] Isin and Ruppert (2020) argue that the COVID-19 pandemic has actually made a new form of government power increasingly visible. They term this register of government 'sensory power' and suggest that it is distinct from the previous regimes of power outlined by Foucault. According to Isin and Ruppert, sensory forms of power have been enabled by the rise of digital monitoring (on- and offline since at least the 1980s). Although the digital architecture of sensory power has been assiduously laid over the last 40 years, it is in the midst of a pandemic crisis that its full potential has been realised. Sensory power is different from Foucauldian governmentality to the extent that it does not seek to govern populations, or the demographic fragments and archetypes identified through censuses and surveys. Instead, it governs through the identification of emerging clusters. During a pandemic, these clusters could be epicentres, hotspots, and bubbles identified by digital contract tracing apps, or the analytical systems of Facebook and Google. During other times, clusters could relate to protest movements, financial transactions, traffic flows, or criminal activity. But a focus on clusters suggests that government is now less about maps and more about digital dashboards (Isin and Ruppert, 2020: 10).

To the extent that it seeks to govern emerging social phenomena, sensory power is closely allied to theories of algorithmic governmentality. But while

---

[2]   Interestingly, Foucault's genealogies of governmental power were in part informed by his observation of the governmental response to public health issues, such as epidemics and pandemics (Isin and Ruppert, 2020).

sensory power is often connected to the predictive capabilities of algorithms, it is also predicated on an historically unprecedent ability to sense existing real-world contacts and movement. Isin and Ruppert's discernment of sensory power within pandemic government uncovers something apparently novel in the history of government. But their analysis is also significant because they reveal multiple systems of governmental power operating on the same problem, but in different ways. Isin and Ruppert recognise how national, territorial, and downright traditional governmental responses were used during the COVID-19 crisis, despite the trans-territorial scope of sensory systems of digital power. While more dynamic in their identification of objects to be governed, smart technologies were often used to monitor and enforce the disciplinary techniques of lockdowns and self-isolation. Often, of course, the use of smart technology appeared to mirror Deleuzian systems of digital control, as individuals were left to moderate their own behaviour in light of the knowledge of nearby COVID clusters and hotspots (Isin and Ruppert, 2020). All of this leads us back to the insight that although smart tech may facilitate novel systems of government, its use by state institutions can equally enable much older systems of power to be realised.

The precise implications of smart technology for emerging forms of governmental power, and associated systems for achieving social order, remain uncertain. Positioning the smart-tech state within the longer history of the rationalities and practices of government does, however, reveal potentially significant forms of historical continuity and discontinuity that are associated with different manifestations of the smart-tech state. In the remainder of this section, we consider the practical manifestations of smart technology within the governmental realm.

Practical Implications of Smart Government

As with many discussions of smart technology, a consideration of the emergence of smart government can tend towards futuristic speculation about what is to come. Although such speculation is important, it is necessary to recognise that smart-tech systems of government are already deeply embedded in our worlds. The fusing of governments and smart technology certainly appears to have popular support among the public. In a survey of 2769 people in 11 countries, researchers at Spain's IE University's Center for the Governance of Change (2021) found support for replacing a proportion of parliamentarians with AI. Fifty-one per cent of the Europeans who were surveyed were in favour of this politico-technological shift,[3] with 60 per cent of respondents supporting the move in America. In China, 70 per cent of those who completed the survey supported replacing politicians with AI! Perhaps Plato's vision of being governed by philosopher kings is going to find its idealised form in smart technology.

The complete replacement of politicians with AI is clearly an unlikely prospect. The cross-cultural support for such a change is most likely connected to people's mistrust of politicians than their warm embrace of smart technology. Nevertheless, even if smart technology is not replacing political systems, it is clearly already an integral part of them. According to Jonsson (Academic Director of the Center for the Governance of Change), few people actually know, or are meaningfully known by, their political representatives (CNBC, 2021). The best a human political representative can hope for is to know the broad needs of their constituency, and to stand for those interests (of course, many politicians may actually be representing other interests altogether). One of the consequences of the numerous intimate interfaces between humans and AI is that smart tech can often discern and predict human desires and interests much more consistently than human representatives can.

The role of smart tech in systems of government goes beyond questions of representation. According to Lago (2021), there are three key ways in which smart technology can contribute to practices of government. First, smart tech can be used to identify the need for governmental intervention (Lago, 2021). The ability to determine the need for intervention can be a short-term product of the sensory power of smart technology (perhaps identifying higher than desired rates of travel during pandemic lockdowns). But as theories of algorithmic governmentality indicate, smart-tech-inspired governmental intervention can also be the product of the inferential constructions of governmental objects of intervention which exceed sensory perception (as may be the case with predictions of terror threats). Second, smart tech can be used to inform the design of government policy (Lago, 2021). Third, smart technology is now central to the evaluation of the efficacy of government interventions. As more of our lives become accessible to digital monitoring, the capacity to assess the effectiveness of acts of government in real time inevitably increases. But smart-tech contributions to the assessment of governmental actions derive not only from enhanced forms of digital surveillance. As we have discussed throughout this volume, smart technology is predicated upon an almost infinite capacity to design, test, redesign, and re-test digital interventions. Related systems are ushering in new forms of experimental government (see Jones

---

[3]   In Spain levels of support reached 66 per cent. But in the UK, 69 per cent of respondents were against the proposal (Center for the Governance of Change, 2021).

and Whitehead, 2018). The nature of machine learning means that the loop between test, design, and re-test requires very limited amounts of human intervention—machine learning can test multiple forms of randomly generated variations of an intervention to identify patterns of optimisation.

As the replacement of politicians with AI raises questions about the nature of politics, the deeper integration of smart technology into the practices of government surfaces questions of ethics and accountability (see Leslie, 2019). Related concerns have been raised about the extent to which smart tech can and should be allowed to infringe on human autonomy at an individual level. In relation to the smart state, these concerns take on a slightly different complexion. When it comes to states (democratic or otherwise), government is inevitably based on individuals ceding certain forms of personal freedom in order to gain collective welfare and security. The continuing legitimacy of states rests on a combined assessment of their relative ability to deliver welfare and security, the ability of the public to scrutinise and assess state practices, and/or the ability of the public to hold governmental officials to account. Much is often made of the fact that the lack of transparency associated with smart technology means that they cannot be scrutinised in the same ways as human representatives. But the paradox of smart government may actually run deeper than this. The optimal performance of smart technology to deliver outcomes appears to be inversely related to the extent to which its processes can be seen and explained (see Lago, 2021). The trade-off between the accountability and effectiveness of systems of smart government is one of a series of issues we explore in the remainder of this chapter as we consider existing smart state systems.

THE SMART NATION—THE RISE OF DIGITAL WELFARE STATES

The provision of welfare services is one of the most significant governmental functions performed by nation states. Despite coming under assault in recent years, initially from the twin forces of neoliberalism and austerity, welfare spending in states such as the UK continues to grow in absolute terms. According to the Office for National Statistics (first published in 2016), in 2017 the UK government spent £264 billion (that is 34 per cent of its total budget) on welfare provision. Welfare spending often covers a broad range of fields, including maternity leave income, pensions, incapacity and disability payments, and unemployment benefits. In addition to the significant financial scope of welfare spending by most states, welfare provision reflects one of the key governmental purposes of the state. Welfare provision reflects the operation of the state as a collective agent that seeks to maintain social order by ensuring that your economic status or capacities do not undermine your right to receive the resources that are needed to live. Welfare is thus a key expression of the pastoral ethos of care which Foucault argued is central to the rationalities and legitimating actions of modern states. However, welfare provision presents governmental challenges to states. The constant need to assess and reassess who qualifies for welfare provision and at what levels places significant calculative burdens on states. The complexities of welfare provision also expose systems to error and fraud. It is in these contexts that welfare has been a prime target of smart state reforms in recent years.

Over the last decade a number of states around the world (including the UK, Australia, and India) have sought to restructure their provision of welfare through the use of digital smart technology (Pilkington, 2019). As with many of the forms of smart technology we have described elsewhere in this chapter, it is argued that the construction of digital welfare states will bring a series of benefits. Pilkington (2019) observes how it is claimed that smart welfare systems will 'speed up benefits payments, increase efficiency and transparency, reduce waste, save money for taxpayers, eradicate human fallibility and prejudice, and ensure that limited resources reach those most in need'.

To realise these benefits, national governments are deploying a mix of biometric identification systems, risk-profiling technologies, algorithms and AI, and a concomitant removal of human caseworkers within the welfare system. While presenting many of the same forms of costs and benefits that are common with smart technology, the development of smart welfare systems raises new, and often troubling, dimensions of the smart-tech society. To explore these issues in greater depth, the remainder of this section considers the establishment of digital welfarism in the UK.

Smart Transformations: Building the Digital Welfare State in the UK

In 2016, the UK government was at the top of the United Nations (UN) E-Government and E-Participation survey (this is a comparative survey of the digital portals offered by states around the world). The UK's leadership in e-government is undoubtedly a product of its drive to become *digital by default*. The UK government's drive to become axiomatically digital can be traced back to its 2012 Government Digital Strategy. This strategy sought to move 'high-volume' government services, such as welfare provision, onto digital portals and platforms. In 2016, the UK state sought to up the ante in its digital crusade through the publication of its Government Transformation Strategy. This strategy envisaged not only the move of government services to digital platforms, but also a 'total transformation' of government through a move to digital forms. The rationale for the British state's digital transformation was explained by Ben Gummer (then Minister for the Cabinet Office), who suggested that the digital state could be more responsive to the needs of its citizens, while also exhibiting greater fairness and equality in the delivery of a complex array of services (UK Government, 2016). The Government Transformation Strategy explicitly notes that shifts towards digital government should also involve the use of algorithms and associated smart technology, which will enable automation in government decision-making (UK Government, 2016).

The forging of a smart government in the UK clearly had the potential to bring benefits to citizens (particularly in terms of the speed and convenience of engagement with different branches of the state). However, this digital transformation cannot be understood outside of its political context. The Conservative Party government who were elected to government in 2010 (as part of a coalition), were strong advocates of austerity in public finances. In this context, smart government can not only be understood as an attempt to develop a more accessible and efficient state system. It was, and continues to be, attractive because it promises significant cost savings. Additionally, this digital revolution of government also offered a form of technological cover for cutbacks in the state's welfare spending. Reflecting on the UK government's digital plans, Alston and van Veen (2019) observe,

> This is presented by the government as an apolitical and technocratic fix aimed at making government more efficient and cost-effective. But in some respects it is also a politicised effort to undermine the social rights of the poorest members of British society, while making it ever more difficult to legally challenge adverse decisions.

Although the smart state system envisaged by the Conservative Government could bring benefits to citizens, it was likely to deliver greater benefits to political elites who could use it to deliver their own societal visions. The potentially uneven distribution of the costs and benefits of smart technology within governmental systems makes discussions of smart government different from our previous analyses of smart technology in the home and workplace. Personal adoption of smart technology may have associated costs, but there remains the (constrained) ability to disconnect from related technologies or switch platforms. There is, of course, less choice about the forms of smart technology that we may have to engage with in the workplace. But should the problems associated with smart workplaces become too great, people may choose to withhold their labour and work for other organisations. As essentially a monopoly provider of governmental services, opting out of the smart-tech state is often not an option (certainly not for the most vulnerable citizens). It is in this context that the rise of smart-tech states requires extra ethical scrutiny.

*

The first government service to go through smart digital transformation in the UK was the welfare system. As part of this transition, the different components of the UK's welfare payments system were brought together under the Universal Credit System. The Universal Credit System harmonised the payment of six streams of welfare payments to low-income households. Although this new welfare system was first announced in 2010 (thus predating the drive towards digital government), many of its features—particularly its emphasis of a single point of welfare calculation—made it an ideal test bed for the experimental deployments of smart state techniques.

The UK's Universal Credit System has already been subject to sustained critique due to the varied deleterious impacts it is having on welfare recipients. The system has been associated with payment deductions, delays in implementation, and significant project overspend (Butler and Walker, 2016). Our primary focus here, however, is not with the implementation challenges or ideological motivations of the overall scheme, but with the lived experience of its smart-tech systems and those forced to live at the Universal Credit System's digital interfaces. According to a UN Human Rights Council report, the Universal Credit System served to place vulnerable people in the UK on the frontline of a digital experiment (UN Human Rights Council, 2019). It is to the experience of these vulnerable groups with the welfare state's new smart-tech interface that we now turn our attention.

Life on the Smart Welfare State Digital Interface

'The British welfare state is gradually disappearing behind a webpage and an algorithm, with significant implications for those living in poverty'.
(UN Human Rights Council, 2019: 13)

One of the most immediate lived experiences of the Universal Credit System has been exclusion. The smart-tech logics of Universal Credit require a digital-as-default approach to the welfare system. It is only by moving wholesale to digital systems that the desired efficiency savings and smart learning potentials can be realised. Reflecting on the Universal Credit System, the UN's Human Rights Council observed,

> The prevailing belief within the Department for Work and Pensions [UK Government] has been that the overwhelming majority of UC [Universal Credit] claimants are online, digitally skilled and confident enough to claim and maintain benefits digitally. In reality, UC has built a digital barrier that obstructs access to benefits, and particularly disadvantages women, older people, people who do not speak English and persons with disabilities. (2019: 13–14)

The reality of the situation was that only an estimated 54 per cent of claimants were able to apply for welfare benefits online without some form of support, with one-third of new Universal Credit claimants failing during the application process (UN Human Rights Council, 2019). Additionally, only *47 per cent* of those on low incomes have access to broadband in the UK (Alston and van Veen, 2019). There are no official estimates of how many people do not even apply for Universal Credit because of these digital demands.

In response to concerns over digital exclusion from the welfare state, it is claimed that offline human support is available for those using the Universal Credit System. But closer inspection suggests that claims of a hybrid human–digital welfare system are limited. Job centres do offer one location where applicants can receive human support, but many have been, or are being, closed (undoubtedly the digital transformation of government is in part driven by the hope that it can reduce the government's need to rent and service office space). The other possible source of digital access and support is public libraries. But once again, libraries have recently seen their budgets reduced and have been closing throughout the UK (UN Human Rights Council, 2019). It is in these contexts that the UN observes a form of digital service outsourcing in the UK:

> The reality is that digital assistance has been outsourced to public libraries and civil society organizations. While library budgets have been severely cut across the country, they have to deal with an influx of UC claimants. Many claimants rely on organizations and charities that are already inadequately funded and under pressure. (UN Human Rights Council, 2019: 14)

In previous chapters we considered the ambiguous interplays between the costs and benefits of smart technology to users. Although smart-tech welfare systems like Universal Credit could, and probably do, bring value to some welfare recipients, something distinctive appears to be happening here. Through the effective denial of service to digitally constrained or disconnected groups, the smart welfare state appears to bring maximal cost and no benefit to some of the most marginalised in society. In this context, the smart state is too smart (or too digital) for core user groups. Such denial of service would be problematic in many areas of life. But as the UN Humans Rights Council (2019) points out, in the UK it is creating a digital barrier between individuals and their social rights.

*

The lived experience of those who have been able to access the Universal Credit System has also been marked by several troubling trends. First, the newly automated welfare system has been characterised by problems of error. These errors have resulted in the underpayment of welfare to some.

In other instances, the smart system has seen the overpayment of benefits. Overpayment has resulted in punitive welfare repayment requirements from recipients and the creation of so-called zombie-debt (Pilkington, 2019). It is, perhaps, to be expected that automatic calculations associated with smart welfare systems would be marked by error in their early deployment phases. It may also be argued that in the long term, smart systems could/should reduce the errors that were previously caused by human oversights or miscalculations. However, there are distinctive problems with the forms of error that are produced within smart governmental systems. Instead of more isolated forms of human error, automated smart systems tend to generate 'error at scale' (UN Human Rights Council, 2019: 14). This error at scale means that problems with under- and overpayment become a more systematic part of the welfare system, at least until the adjustments are made. Error may be unavoidable, but systemic error within the UK's Universal Credit System has seen many of the most financially vulnerable in British society placed in positions of short- and medium-term precarity as errors are resolved or repayments are made.

One of the features of the Universal Credit System is automated systems that detect fraud. While more effective systems of welfare fraud detection are to be welcomed, their current operation means that the digital activity of welfare recipients is likely to be subject to much more intense forms of digital surveillance than that of the general public. The Universal Credit System deploys a risk-based verification process designed to identify which categories of welfare claimants are most likely to be engaged in fraudulent activities (UN Human Rights Council, 2019). The UK's digital welfare state has also deployed a fully automated risk analysis and intelligence system. This system enables the analysis of data on welfare recipients from various governmental databases and social media platforms in order to identify potentially fraudulent actions (UN Human Rights Council, 2019). The ability of smart government systems to operate across public and private digital platforms provides enhanced digital surveillance and learning opportunities. It does, of course, carry with it particular ethical issues. Unlike smart-technology users in the private sector, citizens often do not have the ability to opt out of smart government monitoring systems. It is also not clear that smart government systems are primarily benefiting citizens (it appears in this instance that the state is the primary user beneficiary in the smart welfare system). Ultimately, the UN Human Rights Council (2019) express concern over the deployment of smart fraud-detection systems within the Universal Credit System. They suggest that the system sets a dangerous precedent in relation to the state requiring greater digital engagement with its systems (perhaps your doctor or the National Health Service) so that welfare fraud can be more effectively identified. The UN Human Rights Council (2019) also recognise that the Universal Credit

System brings disproportionate digital scrutiny to the most economically vulnerable in society.

In a 2019 report titled 'COMPUTER SAYS "NO!"', the Child Poverty Action Group reveal the difficulties that welfare claimants encounter when attempting to understand and contest the automated decisions that are produced as part of the Universal Credit System (Howes and Jones, 2019). As a distributed decision-making system, Universal Credit has made it difficult for welfare claimants to understand how their payments are being determined. Of course, critiques of the lack of transparency associated with smart systems and algorithms are well established. Indeed, Amoore (2020) argues that to say that algorithms lack transparency is something of a lazy critique, given that smart systems are constantly having to give an account of their operations. However, we argue that for welfare claimants, the lived experience of smart welfare is deleteriously shaped by their inability to understand the operation of algorithms. This lack of understanding is not, as Amoore argues, necessarily significant at the level of the aggregate operation of the system. This lack of transparency is important, though, in the more specific attempts to make sense of, challenge, and/or live with decisions that are made about specific welfare payments.

The complex nature of the inputs and calculations that inform the Universal Credit System does not mean that it lacks an operational logic. But it does mean the ability of individuals to interpret specific decisions is heavily constrained. In this context, it is perhaps more what smart systems are able to do (namely, draw on more elaborate sources of data and predictive digital inputs), than their mode of operation, which makes them difficult to contest. Amoore (2020) argues that one of the main reasons we should be suspicious of algorithmic governmentality is because it seeks to simplify a complex world in order to make it more governable. Although we agree with Amoore's view when approaching smart systems from the perspective of those who are being governed, we argue that the reverse may sometimes be the case. We claim that the UK government's smart welfare system's problems thus relate not only to the way in which it simplifies complex systems, but also to the way in which it complicates previously simpler ways of doing things.

In their 'COMPUTER SAYS "NO!"' report, the Child Poverty Action Group draw on the testimony of a self-employed father who was claiming Universal Credit. He observes,

> I would say there have been five periods [five one-month assessment periods in UC] at least, which required a lot of untangling. I'm not an accountant, but I'm used to dealing with numbers, I've found it quite incredible the amount of burden that's required to untangle some of these situations I can't possibly see how an average person with less than an accounting degree can deal with all this stuff because it was

complicated for me. For someone who doesn't have free time has kids it's next to impossible to untangle the mistakes. (Howes and Jones, 2019: 14)

Here we see how the complexification (as well as undoubtedly the apparent obfuscation) of the welfare payment system makes it very difficult for those encountering the system to understand how its decisions are arrived at. In terms of living with smart technology, this clearly makes it difficult for individuals to effectively contest payments and correct errors. But, perhaps more importantly, it also introduces much greater uncertainty in relation to life on welfare. The Child Poverty Action Group emphasise how difficult it can be for claimants to not only challenge wrong decisions, but to also understand correct calculations. This means that there is often great uncertainty surrounding how changes in living circumstances may affect their welfare rights. This makes planning and budgeting amid changes in the circumstances of claimants' everyday lives challenging (Howes and Jones, 2019: 12). In terms of the digital welfare state, then, it appears that while smart technology offers significant learning potential and decision-making efficiency for those in government, it actually erodes the learning capacities of citizen users.

Having outlined the use of smart technology in the provision of national systems of government, we now move on to consider the application of smart technology at smaller governmental scales in the context of the creation of smart cities.

THE SMART CITY

The Smart City in Context

This section considers the application of smart technology within urban government. There are clearly overlaps between attempts to govern nations and cities—smart states must inevitably consider how cities fit into wider strategies of territorial government. However, in many nations metropolitan authorities retain significant devolved powers of government, and arguably have more significant impacts on the day-to-day lives of citizens. There is now a noteworthy cluster of Big-Tech companies who are targeting the smart-city market, including Cisco Systems, IBM, Huawei, Verizon, Ericsson, and Alphabet Google's Sidewalks Labs. In 2020 the global smart-city market was valued at $741.6 billion (Strategyr, 2022). Following the COVID-19 pandemic, it is estimated that the value of the market will increase to $2.5 trillion by 2026 (Strategyr, 2022). The rising value of the smart-city market is undoubtedly tied to the promotional boosterism of the Big-Tech sector, but it is not just a supply-side phenomenon. With populations of cities growing rapidly in many parts of the world, the need to act on climate change, and the challenges surrounding resource security (particularly in relation to energy and water), urban areas present increasingly complex and pressing governmental challenges. In this context, smart-city systems offer the promise of technological solutionism to the governmental problems now faced by urban leaders and planners (Mattern, 2016).

There are many definitions of what a smart city is. These definitions tend to emphasise four key characteristics: (1) the use of information and communications technology and smart technologies to improve the delivery of metropolitan services; (2) improvements of the efficiency of urban operations, which reduce energy use, pollution, and the long-term costs of running a city; (3) enhancements in the way in which urban citizens are empowered to communicate with urban authorities; and (4) an ability to better predict and anticipate emerging challenges facing cities. More critical definitions connect smart cities with a new era of urban techno-triumphalism (Mattern, 2016). Mattern observes that smart cities embody a form of 'technoscientific urbanism [that] reflects a neopositivist return to post-war systems' thinking and centralized planning; it is especially visible in the discourse around "smart cities," which regards the intelligence generated from spatial sensing and data analysis as a "fix" for perennial urban problems' (2016: 8). In this context, the smart city appears to have facilitated a renewed confidence within the urban planning profession. There is a belief that, through the development of augmented smart-city intelligence, planners can comprehend the urban process and shape its unwieldy destiny.

According to Kitchin, while smart cities do, in part, embody an ethos of high-modernist planning, they are more a product of the deregulated, neoliberal systems of urban governance that are synonymous with the late 20th century (Kitchin, 2015: 132–3). In this context, while the notion of the smart city does suggest new opportunities for achieving order in the city, it also implies that such order is best achieved through the entrepreneurial zeal of the Big-Tech sector, and not urban planning departments. As a typical expression of techno-liberalism, smart cities do not call for radical forms of governmental intervention, but the use of technical fixes. The smart-city agenda thus seeks to achieve social transformation through the aggregation of marginal gains in organisational efficiency as opposed to more radical socio-economic transformations. Although, as we will see, this is not to say that this is the only direction of travel associated with smart cities.

In the context of this volume's discussion of the smart-tech society, smart cities are significant because they potentially embody an alternative perspective on how we think about the human–smart-tech interface. In one sense, smart cities reflect an upscaling of smart technology from the individual and their smart devices, to collective forms of interactions between urban populations and a smart-urban environment. In this way, smart cities suggest a shift from questions of the quantified self, to quantified communities, and a greater concern with the application of digital technology to collective-action problems. In another sense, smart cities are of interest to us because of their geographical implications. David Harvey (1989) argues that we should think about the city less as a physical object and more as the spatial expression of socio-economic and political processes (for Harvey, these processes are best understood through the overarching logics of capitalist development). If cities reflect the spatial logic of collective human patterns of development, then the smart city could be understood as an attempt to deepen our understanding of these patterns, which at present appear to be escaping our intellectual and practical grasp. On these terms, the smart city should not simply be thought of as just another upscaled expression of the application of smart technology. The smart city could reflect a different order of digital learning wherein smart technology does not only seek to know individuals better than they know themselves but also reveals the nature of the complex socio-spatial systems of which we are all a part.

'Digital to the Core': The Digital City State in Singapore

To explore the experiences of living in a smart city we will focus on the example of Singapore. In some ways Singapore is an obvious choice of focus and may even constitute a somewhat overused case study in discussions of smart urbanism (see Kitchin, 2015). But there are good reasons for considering this nation as an example. The Republic of Singapore is a city state and, as such, embodies an inevitable fusion of smart-city and smart-nation policies. In a 2021 survey, Singapore was identified as the smartest city in the world, beating out competition from Zurich, Oslo, and Taipei (Low, 2021). Singapore's status as a leading proponent of smart urbanism is in part a product of the early start that Singapore made in digitising its urban landscape. At least since the establishment of its National Computerization initiative in 1980, Singapore has consistently placed digital technology at the heart of its development plans. Leading the digital revolution has, it is claimed, been a core strategic element in Singapore's move from the "Third" to the "First" World, and an ongoing basis for establishing international competitive advantage (Yeo, 2022). More recently, smart technology has become central to Singapore's development plans. The vision of creating an intelligent island in Singapore was established in the 2014 Smart Nation programme. This was followed soon after by the 2015 Smart Urban Living initiative and the Smart Government strategy of 2020.

The significance that Singapore's leaders place on smart technology is perhaps best expressed in Singapore's Smart Nation plan of 2018. In this plan the Smart Nation and Digital Government Office liken the digital/smart revolution to the emergence of electricity and the internal combustion engine (Smart Nation and Digital Government Office, 2018). It is clear that in the context of Singapore, visions of a smart urban landscape are not just an aspect of government, they are the basis for Singapore's vision of itself in the future. Although there are various reasons why Singapore has sought to embrace smart technology, it appears that it, in part, relates to a broader sense of national vulnerability. As one senior governmental policy strategist reflected,

> they [political leaders] have a sense of, pervasive sense of vulnerability mixed with a sense of paranoia that if we don't have an excellent public sector, we don't have an excellent government. Why would anyone want to take Singapore seriously? ... all of this is communicated in the public service. Yeah, we are excellent, alert, we are on top of our game. That Singapore's success is entirely ephemeral, entirely man-made. So, we have got to constantly stay ahead of the competition. (Academic and Senior Government Policy Strategist, Public Service Division, Singapore, interview, 2015; quoted in Whitehead et al., 2017: 106)

It appears that Singapore's commitment to smart urbanism is not just a passing fad. The Singaporean authorities see governing through smart tech as being central to securing the competitive economic advantages of the island state. In the deployment of the latest forms of digital intelligence they also perceive an opportunity to legitimate their own existence. This is why it is common in official governmental discourses for Singapore to describe itself as being 'intelligent and digital to the core' (Smart Nation and Digital Government Office, 2018).

The fact that Singaporeans have been living with smart technology for relatively long periods of time, and that many aspects of their urban environments are already saturated with smart devices, means that the city offers a rich context for considering the experience of inhabiting smart-city environments. It is to these experiences that we now turn.

Living in the Smart City

Yeo's (2022) analysis of the application of smart technology in Singapore's Yuhua Estate offers valuable perspectives on the experience of living in a smart city. The Yuhua Estate is located in the western half of Singapore and is comprised of public housing maintained by Singapore's Housing Development Board. Between 2015 and 2018 the Yuhua Estate was fitted with a series of smart cameras and sensors as part of Singapore's Smart Urban Living Initiative (Housing and Development Board, 2014). This initiative covered an area with approximately 9000 residents living in 3194 flats (Yeo, 2022: 8). These smart technologies were fitted in people's homes and in the public spaces of the estate. The Yuhua Estate project was actually comprised of two interconnected projects: the Smart Home initiative and the Smart Neighbourhood programme (Yeo, 2022). The Smart Home initiative deployed smart technologies to facilitate 'elderly monitoring' and more effective energy and water management (Yeo, 2022: 9). The Smart Neighbourhood programme involved the placement of monitoring devices and cameras in public places around the estate (including lift landings and streets) to monitor various estate services, including waste collection and water and electricity consumption (Yeo, 2022: 13). As Yeo notes, the Yuhua Estate has been the centre of a series of experiments in urban planning in Singapore. Public housing estates, where the private and public spaces of the city are owned by the state are, in many ways, obvious places to test out urban smart technology. They offer test beds where it is possible to monitor the urban processes that cross between public and private spaces with relative ease. But this does raise ethical issues, particularly if the same public housing estates are consistently targeted with smart-tech experimentation. While these estates may be the first to receive the benefits that smart technology can bring a city, they are also likely to experience the early-phase problems of smart-tech deployment (Yeo, 2022). Additionally, in collective smart-tech systems the ability of users to opt out of smart-tech programmes is likely to be much more constrained than at an individual level. If lower-income residents are being consistently targeted by urban smart-tech experiments—which may not be optimised, and only offer limited opt-out potential—then smart cities could quickly embed systems of spatio-technological injustice.

Yeo (2022) surveyed residents about their experiences with smart-home devices. A common response from residents about the initiative was disinterest. Yeo speculates that this disinterest could be a product of the fact that smart technology is actually quite marginal to the everyday private lives of residents in Yuhua Estate. This, of course, may speak to precisely who is benefiting most from the learning loops of the smart city. It could be that unlike the individual adoption of smart tech, the collective provision of smart technology is always focused more on what can be collectively learned about urban life by the governing class. Yeo also suggests that residents' indifference to the presence of smart tech is the product of the normalisation of smart technology in a place like Singapore (or its entering a form of technology unconsciousness; see Beer, 2019). Of course, the normalisation of smart tech as an unnoticed aspect of urban life has benefits for the learning capacities of the smart city. The less that people are aware of smart technology, the more likely their behaviour 'in the wild', so to speak, will reflect an unfiltered reflection of their habits and customs (see Zuboff, 2019).

There was evidence of some resistance to the insertion of smart technology into people's homes on the Yuhua Estate. According to Yeo, resistance was particularly evident among more elderly residents who saw the presence of smart technology as too disruptive and its operations too difficult to learn (2022: 12). Resistance to smart-city technology among Yuhua's older residents is significant, given that the use of smart technology to support Singapore's ageing population is a central goal of its digital strategies. Singapore's Elderly Monitoring System utilises smart motion sensors, which are able to discern routine patterns of daily movements among older residents and trigger alarms when these patterns change, or no motion is detected at all (Smart Nation Singapore, 2022). Although these forms of intrusive technological surveillance have raised ethical concerns, it is argued that they are enabling elderly residents to continue to live independent lives; it also potentially means that less strain is placed on health and social care services (World Economic Forum, 2020b). In many ways Singapore's Elderly Monitoring System is a prototypical example of smart-city governance, combining as it does smart-monitoring and pattern-recognition technologies within broader systems of collective social and health care support. It also embodies the inherent ambiguities of these technologies: on the one hand, they promise greater independence and empowerment to elderly relatives, but on the other hand, they are an unwanted and potentially confusing technological disruption within the home.

Yeo's research also reveals that some Yuhua residents were not engaging in the scheme because they were unconvinced that smart technology would bring significant benefits. There appears to be a sense that the operation of homes (in terms of energy and water use at least) was already 'good enough' (2022: 11). These sentiments appear to relate to the Utilities Management Systems that are now being rolled out across Singapore. These systems monitor water and energy use patterns around the house and enable devices to be controlled remotely in order to save energy. The notion of being 'good enough' reveals a form of existential scepticism among residents about just how much optimisation smart technology can bring. This perspective may speak to a deeper issue at the heart of smart government concerning the relation between claimed efficiency gains in citizens' actions and the inconveniences that the continual pursuit of marginal gains brings. In this context, while smart-city technologies may support the delivery of aggregate (and significant) efficiency gains across a population, the inconveniences they generate at an individual level may outweigh the benefits they bring at that scale.

Yeo's research also considered the social experience of smart-city technology in the public spaces of the Yuhua Estate neighbourhood. The Smart Neighbourhood programme involves the deployment of a combination of smart cameras and monitors throughout the estate. Yeo reveals that there were varied social responses to the neighbourhood's smart upgrades. As with the smart-home initiative, there were some on the estate who were ignorant of or indifferent to the upgrades. Of those who did note a change in experience of the neighbourhood, some residents suggested that it generally made them feel more secure (Yeo, 2022). Interestingly, this sense of enhanced security led some residents to also feel free to be able to move around the neighbourhood safely (Yeo, 2022: 12). While some residents reported enhancements in their behavioural freedoms, others claimed that neighbourhood smart tech restricted their sense of freedom. In the presence of new digital surveillance monitors and cameras, Yeo reports that some of Yuhua's residents felt the need to moderate and monitor their behaviours to ensure that they were appropriate for public spaces (2022: 12). In this context, Yeo draws connections between Yuhua's residents' modified behaviours in the smart neighbourhood and Vanolo's (2014) notion of *smartmentality*. According to Vanolo, the use of smart surveillance technology in public places tends to produce docile subjects who self-regulate their behaviour to conform with socio-governmental expectations. One may question the extent to which awareness of specifically 'smart' monitoring is what causes behavioural modification here, or if it reflects the forms of self-regulation that are typical of much older forms of analogue surveillance. It seems to us that the notion of smartmentality, if it does exist, is more appropriately applied to situations when the behaviours of citizens are shaped on the basis of the intelligence that is gained from urban smart tech, and not just because of its presence in the landscape.

*

Yeo's analysis reveals that people's relationship with smart-city technology is often different to more individualised engagements we have discussed earlier in this volume. Their analysis also exposes very different understandings and responses to the deployment of digital tools of government in their homes and public spaces. While some people remain indifferent, or suspicious of publicly provisioned smart tech, others appear to embrace its liberating promises.

CONCLUSION

Smart technology is clearly becoming an important tool within the emerging systems of government evident at various scales throughout the world. Japan is arguably now leading this technological revolution in the application of smart technology through its national vision of Japan 5.0, 'a cyber-physical society in which ... citizens' daily lives will be enhanced through increasingly close collaboration with artificially intelligent systems' (Gladden, 2019: 1). Some argue that smart-tech government reflects a distinct break in the history of liberal society, with the emergence of a more predictive and anticipatory system of government that is able to actively shape behaviour at a large scale. Others claim that smart-tech government is actually the fulfilment of a much older governmentality in which states seek to monitor and protect each and all. Within this chapter we have also seen how the governmental deployment of smart technology can result in a new division of learning within which the primary benefits of digital monitoring are accrued by governmental elites and not those subject to surveillance (Zuboff, 2019). We have also charted how, in the context of both the digital welfare state and smart cities, an individual's choice of whether to engage with smart technology is much more heavily constrained.

A series of authors have started to question the broader implications of smart technology for citizenship. Mattern's (2013, 2016) pioneering analysis suggests that smart cities embody a change in the epistemological strategies that are used to understand and intervene within public life. Mattern argues that within the smart city there is a tendency to prioritise quantitative ways of comprehending city life ahead of more qualitative human perspectives on urban living. This epistemological bias tends to result in the marginalisation of significant forms of collective and individual intelligence (Mattern, 2016). A focus on the quantitative data signals emitted by citizens, as opposed to qualitative insights, reflects a shift in the nature of political engagement with government. According to Mattern, 'In this universe, citizens relate to their city by consuming and administering its systems, and by serving as sources of measurable behavioral data' (2016: 10). This system inevitably prioritises a distinctly passive form of citizenship, where human input into governing systems is reduced to the automatic provision of user data (Mattern, 2016: 13; see also Gabrys, 2014). There is, of course, no reason why smart governmental systems should preclude more engaged and engaging forms of human input. But the epistemological orthodoxies of many smart-tech projects do not appear to be particularly welcoming to them. It is in this context that the rise of smart governments reflects more than just new ways of governing; they challenge established democratic norms of political engagement.

8. Dumbing down—recalibrating our relations with smart technology

An emerging feature of the smart-tech society are groups of people who are, in various ways, disconnecting from or reorienting how they connect with digital technology. Related developments include slow computing, digital veganism, Facebook deleters, de-platforming, and even the use of smart lock-out apps. These processes of disconnection and reorientation remain fairly marginal developments compared to the increasing number of people who are deepening and extending their connections with smart tech. They do, however, reflect an emerging trend which, we argue, has important implications for future social relations with smart tech. In this chapter we use the term 'dumbing down' to characterise these emerging practices.

Dumbing down refers to two distinct but interconnected processes. First, it reflects deliberate attempts to de-smart our lives by disconnecting (or connecting more carefully) with digital technology. On these terms, dumbing down may result in less dependency on technology as a source of transactive memory and decision-making. Second, dumbing down refers to the costs of reorienting our relations with smart tech. These costs can take different forms. They may relate to the social costs of becoming digitally disconnected from online communities. The more troubling costs may relate to the relative loss of cognitive capacity that results from being cut off from the deep learning of our attendant smart devices and algorithms. With the social and cognitive costs of disconnection being potentially high, it is important to ask what is driving people to reorder their relations with smart technology. Explanations for such trends are, of course, not hard to come by. Political scandals, including the Cambridge Analytica affair, the Snowden revelations, Facebook's emotional contagion trial, and Mark Zuckerberg's secretive meetings with Donald Trump, are regularly used to explain digital disconnection. Digital disconnection is also routinely connected to the addictive qualities and mental health consequences of smart tech. From the personal to the political, this chapter argues that the reasons for people reorienting their relations with smart tech are more complex and socially nuanced than these explanations would suggest. This chapter begins with a consideration of the different ways in which forms of digital disconnection are manifest and the existing theories that can help us understand these trends. These discussions are positioned within a broader historical consideration of social resistance to technologies of different kinds. This chapter primarily focuses its attention on the emerging community of Facebook deleters. Drawing on in-depth interviews with members of this community, the analysis considers the processes that are associated with deletion (as it happens, it is rarely a case of simply being on or off social media). We also reflect upon the motivations that inform the actions of deleters and the consequences of their actions.

DIGITAL DISCONNECTION: TECHNOFERENCE AND BLOCKING SMART TECH

Why We May Want to Block Smart Technology and Why We Can't

Perhaps the most obvious, and certainly the most newsworthy, way of reorienting our digital lives is by disconnecting from smart technology. There is no shortage of former Big-Tech innovators and leaders who are now encouraging us to abandon the very technology that they helped to build. Prominent books, such as Jaron Lanier's (2018) *Ten Arguments for Deleting Your Social Media Accounts Right Now*, emphasise the benefits of digital disengagement (see also McNamee, 2020). The reasons we may choose to disconnect are well known. They range from health issues (addiction, stress, mental illness) to more societal questions of behavioural manipulation, privacy, and the loss of physical connection with others. Many of these arguments centre around the idea that smart technology is undermining our ability to be human at both an individual and collective level. For example, in their reflections on the impacts of smart tech on human hormones (specifically the exploitation of dopamine rewards and the potential suppression of oxytocin), Rashid and Kenner observe:

> The shifting balance of our hormonal drivers towards online activities combined with the resulting loss of attention to the 'real' world and the lessening of physically present social bonding could have grave consequences. It is not hard to see how these trends could irreversibly damage future generations' ability to commit fully and deeply with others in the real world. (2019: 71)

McDaniel and Coyne (2016) describe the disruptive influence of digital tech on our lives as *technoference*. The notion of technoference tends to be applied to specific instances where smart tech distracts us from real human connections (particularly with our friends and family). The term could, however, be used to depict the more general disruptive influence of smart tech on our ability to be our pre-digital-tech selves.

While these justifications for drawing back from smart tech are important, we believe that they greatly simplify a complex set of processes. The idea that there is a real world that smart tech cuts us off from is clearly problematic. It is clear that our 'real' lives are already complex amalgams of analogue and digital existence. Our digital lives facilitate and structure new in-person experiences, just as our real worlds influence our digital lives. Furthermore, the figure of the cyborg, to which we have consistently alluded, reminds us that there is no pre-technological primordial human state to which we can realistically return. Even if such a state of pre-smart-tech nature did exist, we probably wouldn't actually want to go there, at least not for long.

All of this is to say that many contemporary discourses of smart-tech (and digital) deletion and liberation are at best simplistic and at worst wilfully unrealistic. They take a complex social problem and suggest an immediate solution. To be fair, those who promote digital disconnection and unplugging recognise some of the complexities of what they are calling for. Lanier, for example, acknowledges that 'being able to quit social media is a privilege' that is not available to everyone (2018: 3). But he still claims that if you can quit, you should, because '[q]uitting is the only way, for now, to learn what can replace our grand mistake' (2018: 3). But encouraging digital elites to venture into the smart-tech-free world and playfully explore how to build alternative futures is not a great source of help for everyday life in the digital and now. But this is more than just an issue of differential capacities of disconnection. We already appear to be in a situation, in many parts of the world, where a full block on smart tech is an impossibility for pretty much everyone.

Hill's Block

We preface our discussions of smart-tech disconnection and recalibration with a recognition that full disconnection from the smart-tech society is already something of a practical impossibility for many. In 2019, New York-based tech journalist Kashmir Hill conducted an experiment in smart-tech disconnection. With the logistical support of the tech news site Gizmodo (it turns out you need a lot of logistical support to even attempt a full block), Hill attempted to cut the so-called Big Five digital-tech giants (that is, Amazon, Google, Facebook, Apple, and Microsoft) out of her life over a six-week period. It is worth reflecting on this salutary experiment at some length as it offers valuable context for the stories of smart-tech disconnection and digital reorientation which follow in the chapter.

Hill's experiment was a reaction to the response which many people who express concern over Big-Tech receive: 'If you don't like them, just stop using their services' (see Hill, 2019a). This is a classic neoliberal rejoinder to corporate problems: if you disapprove of something, use your consumer power to send a signal to that company, and the market will do the rest. As with so many neoliberal aphorisms, however, this suggestion shows scant appreciation of the working of contemporary monopolies. When it comes more specifically to the smart-tech sector, this response appears to be totally ignorant of the cyborg-like connections that already fuse many to digital technology. The idea of disconnecting from smart tech also has broader significance. For many it serves the function as a kind of last refuge of human agency: yes, I am deeply dependent on smart tech, but I control the kill switch. Hill's experiment is thus perhaps best conceived of as an attempt to test these theories (perhaps ideologies is actually a better word here) of human control over smart tech.

Hill's experiment was an iterative one. Each week, for five weeks, she disconnected from one of the Big Five tech giants. Then in week six of the experiment she attempted to disconnect from all five, all at once. A sense of the scale of this experiment can be gained from Hill's first report on her trial, published on 22 January 2019:

> It's not just logging off of Facebook; it's logging off the countless websites that use Facebook to log in. It's not just using DuckDuckGo instead of Google search; it's abandoning my email, switching browsers, giving up a smartphone, and living life without mapping apps. It's not just refusing to buy toilet paper on Amazon.com; it's being blocked from reading giant swaths of the internet that are hosted on Amazon servers, giving up websites and apps that I didn't previously know were connected to the biggest internet giant of them all. (Hill, 2019a)

The challenges of disconnecting start at a very practical level. Our everyday experience of smart tech is through its physical manifestation in our daily lives in the form of phones, digital home assistants, and laptops. It is, however, the Big-Tech servers which arguably constitute the most important human interface with smart technology. To block connection with these powerful, yet inevitably opaque, servers, Hill had to call on the services of a tech expert to build a bespoke virtual private network (VPN) to channel internet traffic and avoid the IP addresses of the Big Five. Interestingly, it was only once those servers went dark that Hill was able to perceive the infrastructure of the smart-tech society.

During her first weeks of digital disconnection, Hill stopped using and blocked Amazon servers (Hill, 2019b). During one week of her smart-tech block, Hill discovered that her devices attempted to contact Amazon servers nearly 300 000 times! This high rate of attempted contact is a product of the fact that while we primarily think of Amazon in terms of its retail services, the majority of the company's profits come from Amazon Web Services, which provide the digital infrastructure for a broad range of companies. Disconnecting from Amazon thus meant Hill inadvertently blocked 23 million IP addresses (including the US Government Accountability Office website), various apps, and the workplace communication platform Slack. It is interesting to think that it is only by blocking smart technology that we are able to see what it actually is. While normally hiding in plain sight, when blocked, Amazon's umbra became apparent to Hill.

Throughout this volume we have explored the human experience of living with smart technology. An important part of this experience is, of course, what happens when we try to live without smart technology. In this context, what is most interesting about Hill's block of Amazon is its impacts on her family and social life. Deciding to disconnect from smart technology is not something we can necessarily do in isolation: it has ripple effects for our family and wider social circle. In this context, it is interesting that the low point in Hill's Amazon block came when she found herself consoling her crying one-year-old daughter because she could no longer watch her favourite TV shows (Netflix was taken out by the block because it also uses Amazon Web Services). Our dependence on smart technology is in part secured by our dependents' dependence on its associated platforms.

During her second week of her smart-tech blockade, Hill attempted to disconnect from Facebook (we will reflect on this process in our discussion of Facebook deleters later in this chapter). The next tech giant to be blocked by Hill was Google. The process of de-Googling had significant impacts on how Hill experienced the internet and her ability to navigate everyday life in the real world (Hill, 2019c). The most immediate impact was on her working life, however. As with many people who don't have the benefit of working for corporations with software licensing agreements with Microsoft, Google offers a freely available set of software options (Google Calendar, Gmail, Google Docs, Google Chrome, etc.). Thus, de-Googling did not just have implications for Hill's social life and consumption activities; it made doing her job far more challenging. The prospect of deleting Gmail also made Hill realise how 'the infinite space offered by the tech giants has made us all digital hoarders' (Hill, 2019c). The notion of digital hoarding is different from the form of curated content you might expect to find on social media platforms. It is a more random collection of contacts, attached documents, and useful links. This data hoard is, of course, useful to our digital transactive memory (I don't need to remember that because Gmail will do it for me). But it is also of clear value as a data source for the predictive analytics of Google.

Perhaps the most intriguing aspects of Hill's de-Googling was the impact it had on her experience of the internet. As Hill's VPN blocked Google addresses, she noticed a clear slowing down in the operation of the internet. It turns out that this slowing down was not the product of her attempt to reach Google addresses that were being blocked. The slowing-down effect was the result of non-Google sites Hill was visiting unsuccessfully trying to contact Google. It turns out that before providing us with the content that we require, sites first seek to serve the needs of Google's trackers, ads, and analytics (Hill, 2019c). Searching unsuccessfully for Google thus slows these sites down and makes our experience of the internet much more challenging. Here again we see how it is only in severing our connections with smart tech that the nature of its work and infrastructure become apparent to us phenomenologically.

During Weeks 4 and 5 of her digital blockade, Hill sought to remove Microsoft and Apple from her life. Hill found that cutting out Microsoft was the easiest block to make. It did mean that she could not use her in-car entertainment system, but as an Apple user, Hill was able to carry on life pretty much as normal without Microsoft (Hill, 2019d). The one area where Hill could not completely cut Microsoft out of her life was in relation to digital payment in shops. Microsoft supports monetary transactions payments in a wide variety of retailers. Not only could Hill not control this, but she found that she was often unaware of which shops and facilities use Microsoft systems and which do not. This issue reveals a further set of limits on our ability to regulate our relations with smart-tech platforms. We may attempt to cut them out of our day-to-day lives, but we cannot control the digital systems that others use and we may rely on.

Given her reliance on Apple products, Hill's withdrawal from using its services was much more challenging than her experience with Microsoft. But Hill's so-called *iLoss* was also very different from her block of Amazon, Facebook, and Google. Apple does not rely on the forms of surveillance revenues associated with Facebook and Google; as such, its connections with consumers are mediated primarily through its devices and not its data-gathering infrastructure. Thus, if you don't use Apple devices, you won't technically need to block its services (you will already be on the outside of its digital walled garden; Hill, 2019e). But herein lies the particular challenges of *iLoss* for Apple users. If you use Apple, you are most likely to be locked into its digital infrastructure. Thus, blocking Apple is not just a matter of changing your web browser, but involves the expensive retooling of your smart-tech life. Hill describes this challenge in the following terms:

> Apple is my gateway to almost all things digital. I am physically touching an Apple device for the majority of any given day. Being asked to remove Apple from my life was like being asked to remove a part of my body that was incredibly useful but that I could live without, like a finger or an eyeball. (Hill, 2019e)

Hill likens her *iLoss* to an actual eye loss! This of course echoes our discussion of the connections between smart tech and the figure of the cyborg (see Chapter 2). To these ends, blocking smart tech is not just a digital process, but a corporal one. With the likely uptake of smart-tech implants in the future, the price of this physical disconnection could be surgical.

The denouement of Hill's experiment came in Week 6. This is the week in which she had to block all five of the Big-Tech giants (see Hill, 2019f). As you can imagine, this week reflected an amplification of the previously discussed impacts of blocking each of the tech giants individually. What surprised Hill during the week of her full-tech block is that blocking smart tech essentially involves giving up on all digital technology. Due to the interoperable nature of so much digital technology, abandoning the smart-tech giants meant that Hill was unable to use all but the most basic digital technology (such as a Nokia 3310 mobile phone; Hill, 2019f). This is essentially why disconnecting from smart technology is now already an impossibility for most people. It is not just that you must disconnect from the data industrial complex. Because the smart-tech industry is so embedded in our entire digital infrastructure, there are effectively no 'dumb' digital alternatives to fall back on. So, just like Thoreau's retreat to Walden Pond, to go dark in the smart-tech society would involve a form of withdrawal from socio-economic life that is simply not possible for most people. With significant technical help, Hill was able to achieve a full smart-tech block for a limited time. What she experienced during this time is instructive. Hill describes how her life became significantly quieter (she did not have access to YouTube, Apple Music, her Echo, Netflix, or Spotify). She also noticed how her mind tended to wander more, opening up new avenues for creative speculation. In a video produced in association with her experiment, Hill also reflected upon how her tech block meant that she was engaging with her daughter in new ways and giving her more undivided attention in the morning. What is palpably clear from this experiment is that while blocking smart tech may be very difficult, doing so opens up new (or perhaps we should say old) ways of being human and of sustaining more 'authentic' forms of human experience.

There is a series of key conceptual insights that emerge from Hill's groundbreaking experiment. The first is the fact that our relationship with smart tech must be understood as a collective-action problem.[1] In this context, our relationships with smart technology are less about individual choice and more about our collective use and investment in related technologies. This is an issue that is central to this volume and the very notion of the smart-tech society it explores. The key issue here is that while we experience smart tech and its outputs at an individual level, its power and influence operate at the social level. Changing our individual relations with smart tech can thus only meaningfully be achieved at the collective level—unless we are willing to individually forfeit our right to meaningful participation in society. The second is the issue of convenience. It appears that changing our (individual) relations with smart technology is primarily a matter of convenience. The most

---

[1]   This issue is raised in an interview that Hill conducted with Daniel Kahn Gillmor (a technologist at the American Civil Liberties Union).

immediate impact of blocking smart tech appears to be to deliberately inconvenience ourselves. This inconvenience operates at two main levels. There
is the loss of the expediency that comes from using the smart learning that is
built into smart-tech systems (fast Google searches, personalised Amazon recommendations, etc.). We could, perhaps, call this shallow inconvenience due
to its relatively minor impact on our daily lives. But there is also a deeper and
more consequential inconvenience associated with blocking smart technology.
This is an inconvenience that derives from cutting ourselves off from the data
industrial complex and its monopolistic control of so many of the collective
interfaces that make our society function. This is where the monopolistic
logics of the smart-tech society really start to bite deep into our choices and
capacities as human beings. To remain competitive, smart-tech giants need to
be able to gather as much of our personal data as they can. This requires the
control of as many digital spheres as possible. Smart-tech monopolies are thus
problematic not just because they prevent meaningful competition in terms of
service delivery, but also because they inevitably deny us opportunities to be
human in different ways.

Ultimately, while Hill's tech block demonstrated that disconnecting from
smart tech was a practical impossibility for her in the long term, it did inspire
her to change her relations with digital devices and platforms and to become
more aware of her digital wellbeing. In his book *We Are Data*, Cheney-Lippold
describes how the data industrial complex often leaves us with a 'sliver of
agency' (2017). While Hill was able to temporarily expand her agency in
extremis, in the following section we consider what others are doing to maximise this trace of autonomy through a series of adaptive digital practices.

REORIENTING DIGITAL LIVES: SLOW COMPUTING

Hill's experiments reveal the impossibility (for many) of abandoning smart
tech entirely. The question remains though as to what else we can do to safe-
guard ourselves against the problems of smart tech, while still experiencing
its benefits. In this section we explore a series of evolving movements and
strategies that, while stopping short of smart-tech abstinence, offer alternative
frameworks for human–smart-tech interaction. While varied, these movements
and strategies can be usefully categorised as different versions of slow computing (see Kitchin and Fraser, 2020). In many ways Hill's block can be thought of
as a form of digital veganism—a principled rejection of surveillance-oriented,
closed smart-tech systems and the pursuit of more ethical digital alternatives.
The phrase digital veganism was coined by start-up founder Cody Brown as
a pejorative term to denote the idealistic virtue-signalling associated with those
who prefer open-source technological alternatives (Jeffries, 2011). Although
slow computing can be motivated by similar ideals as digital veganism, it

reflects a much broader set of pragmatic and adaptive responses to smart technology.

The term slow computing was first deployed by Nathan Schneider in an article he wrote for *The New Republic* in 2015. In the wake of Edward Snowden's revelations concerning the pernicious global surveillance practices of intelligence agencies, Schneider (2015) transitioned to using open-source software and cloud services operated by a 'democratic membership organiza-tion'. There are two salient aspects to Schneider's transition away from using Big-Tech: first, it was not easy, and, as we would expect, it inconvenienced his life; second, and perhaps more surprising, Schneider did not find this inconvenience frustrating, but actually associated it with certain pleasures. The pleasures associated with Schneider's (2015) transition led him to liken his experience to the emerging Slow Food Movement: 'As in Slow Food—with its unhygienic soil, disorderly farmers' markets, and inconvenient seasons—the annoyances of slow computing have become pleasures. With community-made software, there's no one to blame but us, the community'.

There seems to be something in the disorderly nature of slow comput-ing, with its community focus (Schneider was able to visit the servers that supported his cloud service), that is appealing to Schneider. Perhaps this is because slow computing feels like it is closer to more traditional forms of human experience. Crucially, in his articulation of slow computing, Schneider appears to identify a zone where inconvenience can enhance, as opposed to erode, the quality of human experience, and where digital intimacy and a sense of community trump scale and efficiency.

Beyond the reference to organic agriculture and bucolic ideals, Schneider also identifies a political dynamic within slow computing. Schneider (2015) thus compares the democratic claims of slow computing to those of Big-Tech:

> There is a habit in tech culture of saying that the latest app is 'democratizing' what-ever it happens to do. This is lovely, but best not to confuse it with actual democ-racy. Democracy is about participation with control, freedom with accountability, privacy with transparency. Free-and-open software, however, operates on a differ-ent time-scale. Since nobody owns it, it's harder to become fabulously wealthy from it. People make these programs because they need them, not because they think they can manipulate someone to want them.

For Schneider, then, slow computing is political because the move to open-source software enables users to know more about the digital services they are using and to hold their architects accountable. But Schneider's vision of smart computing is constrained by the very thing in which he finds libera-tion. The emphasis he places on alternative, open-source platforms makes slow computing feel technologically unattainable (and perhaps even undesirable) for many. The future can only be free if it involves Linux.

Within the work of Kitchin and Fraser (2020), slow computing becomes a broader project of socio-digital reorientation. Building on the formative ideas of Schneider, Kitchin and Fraser develop the principles of slow computing into an amalgam of political goals and a wellbeing project. For Kitchin and Fraser, slow computing is thus, 'about seeking an alternative path to the speed and busyness of the modern world that prioritises a different set of values—enjoyment, patience, individual and collective wellbeing, sovereignty, authenticity, responsibility, and sustainability' (2020: 12). This vision of slow computing is of interest to us because of the way it acknowledges both the joys and problems of digital life and attempts to navigate a realistic path through these tensions.

According to Kitchin and Fraser, the move towards slow computing is a response to two broad sets of processes: those of acceleration and extraction. The notion of acceleration speaks to the tension between the convenience and controlling aspects of the smart technologies we have been exploring throughout this volume. The acceleration of social life within the smart-tech society is described as, 'a sense that, for all the convenience of digital technologies, they also cut into our lives in ways that increase the pressure to always do more, always remain connected, always stay alert, leaving us struggling—at times, really struggling—to switch off, relax, and not reconnect' (Kitchin and Fraser, 2020: 10).

This quote is interesting because it speaks to the social costs of smart technology: for always-on technologies to reach their potential, they require that humans are also always on. The notion of always-on-humanity is suggestive of the ways in which smart tech depends on the digital sharing of data and the associated human labour this involves. It is also suggestive of the ways in which, in addition to offering labour-saving opportunities, smart tech often requires humans to match the productive capacity of their digital-tech counterparts (Berreby, 2021a). According to Beer (2019), the acceleration of social life is aligned to a rationality of speed that suffuses the world of big data analytics. This rationality of speed suggests the temporal urgency of deploying and enhancing data analytics in order to cope with an accelerating social world (Beer, 2019)—and so the circular logic of acceleration within the smart-tech society is complete!

Extraction, of course, refers to the ways in which we exchange data on various aspects of our social life and our preference for 'free' use of smart-tech platforms and services (Kitchin and Fraser, 2020: 11). In keeping with established theories of surveillance capitalism, Kitchin and Fraser are concerned with the practices of extraction because of the ways in which it makes the producers of digital data (us) the targets of tracking and behavioural manipulation by often unregulated tech giants (2020: 11). According to Kitchin and Fraser, the combined, mutually reinforcing practices of acceleration and extraction generate a form of pressure that is becoming endemic in the smart-tech society. But they also emphasise the choices that are available to people in the middle of this smart-tech maelstrom:

> we are struck by the way individual users—you and I—are in the centre of things with serious pressures bearing down from two directions. But we also know that individuals can choose, to varying degrees, how they act in response to acceleration and extraction, and can also work with others to counter these pressures. (2020: 10)

For Kitchin and Fraser the processes of acceleration and extraction invite forms of human response (2020: 11). Slow computing is then best thought of as a collection of responses to the smart-tech society that lie somewhere between Hill's technological block and an unthinking surrender to the demands of smart tech.

In practical terms, the slow computing envisaged by Kitchin and Fraser involves a series of life hacks that serve both to inconvenience our lives and to liberate them (2020: 77). These practices are organised under the broad regimes of time and data sovereignty. As the name suggests, establishing time sovereignty involves working against the forms of hyper-communication and hyper-coordination produced by digital devices. Kitchin and Fraser suggest that the demands of digital communication and coordination shift humans into a form of (unthinking) behavioural response (see Chapter 4, this volume). This tends to mean that while our attention is captured by smart tech, we pay relatively little attention to what we are actually doing on our smart devices. Smart computing encourages people to regain control of their time by reasserting attention onto what we are actually doing in our digital lives. Kitchin and Fraser argue that 'a slow computing day requires deliberate contemplation about where, when, and how you participate in chains of responses' (2020: 78), and the conscious breaking of those responsive patterns where and when they undermine our ability to manage our own time and wellbeing. Practically this may involve only looking at your emails at certain times of the day or switching off your mobile phone more regularly. But establishing time sovereignty is *sine qua non* with slow computing: if we cannot take control of how we use time, it is very difficult to break the unthinking responsive states that are associated with smart technology.

According to Kitchin and Fraser, the practices of data sovereignty involve the deliberative curation and withholding of digital data about our lifestyles and preferences. The establishment of data sovereignty can take various forms, but at its core it is a rejection of the economic models that are associated with Big-Tech. A key practice in establishing data sovereignty is the act of 'stepping off' surveillance capitalist platforms, such as Facebook and Google. Much has been said about the differential ability of people to actually do this—with such options seeming to be more available to the affluent than the poorest in society (Lanier, 2018). But it is also argued that if this is not an option for all, it is still important that those who can do step off, if only so that they are not actively supporting an exploitative business model (Lanier, 2018). In keeping with Schneider, establishing data sovereignty is also associated with the use of open-source online platforms and systems, including Linux operating systems and the Tor browser. It is of course here that we again see the connections between slow computing and an embracing of digital inconvenience. It is interesting to note that inconvenience may not simply be an inevitable cost of reestablishing autonomy in the smart-tech society. As Schneider intimates, inconvenience (as found in his farmers' market analogy) appears to be conducive to certain aspects of the human condition. This is, of course, why human communication is not just about relaying information, or why we may not always want to take the most optimal route home as specified by our sat navs (Munger, 2021). Convenience does not provide everything that we need socially (including reassurance, surprise, stimulation, and distraction). Indeed, there appears to be a strong connection between inconvenience (broadly defined) and human wellbeing and longevity (Buettner, 2012).

Data sovereignty can also be achieved through the tactics of obfuscation (Kitchin and Fraser, 2020). Kitchin and Fraser describe obfuscation as 'the deliberate use of ambiguous, confusing, or misleading information to interfere with data extraction' (2020: 96). These practices not only serve to reduce the amount of data being collected by smart technologies (as with acts of stepping off), but also to actively disrupt the data which related devices and platforms receive. While we recognise the disruptive potential of obfuscation, as a technique of digital self-reorientation it has some problematic aspects. As with stepping off, effective forms of obfuscation are likely to be time-consuming and inconvenient. Perhaps more troubling is the way in which they seek to make deliberate forms of dishonesty a form of human virtue. We do not have an ethical problem (necessarily) with deliberately duping surveillance capitalists. The idea of promoting deceit as a human virtue does, however, strike us as an unwelcome outcome of social interactions with smart technology.

While offering a fairly holistic roadmap for digital self-reorientation, Kitchin and Fraser (2020) do acknowledge shortcomings in their vision of slow computing. They acknowledge that at an individual level it involves working against powerful forces of persuasive habituation that are associated with smart technology. They also acknowledge that to be effective in the longer term, slow computing will need collective-action from the state, civil, and corporate sectors. It is only through such forms of collective-action that the socio-economic obligations and expectations that fuel individual engagement with digital tech can be addressed. But, as Kitchin and Fraser acknowledge, at the level of collective-action there are still powerful ideological processes that serve to promote and normalise the practical forces of acceleration, abstraction, and smart-tech engagement.

*

The remainder of this chapter offers an empirical analysis of people's attempts to reorient their relations with smart technology. The next section reports on a SenseMaker survey of 120 smart-tech users (see Chapter 3 for more about this survey). This survey considers the primary motivations that influence people's attempts to block and reorient their relations with smart tech, and the ease with which these changes can be made. The final section draws upon in-depth interviews with Facebook deleters. Together these sections offer empirical data that build on and extend Hill's auto-ethnography and consider what may drive and constrain forms of slower computing in the wild.

UNDERSTANDING DIGITAL REORIENTATION: MOTIVATIONS AND CHALLENGES

In Chapter 3 we introduced a SenseMaker survey designed to assess people's responses to smart tech's predictive interventions within their lives. The results of the survey provided us with some potentially valuable insights into individual motivations and capacities for reorienting their relationships with smart technology. While it is important to note that these results are in part a response to reflections on one particular social media platform's predictive interventions into people's everyday lives, they do offer more general insights into participants' relationships with and perceptions of smart tech (we reflect here, in a slightly different context, on some of the same data that we introduced in Chapter 3).

It is interesting to note that few people in our survey described their chosen social media platform as a necessity (Figure 8.1). As we have previously discussed, however, the importance of smart-tech platforms is often not immediately apparent until you stop using them. Nevertheless, there was a cluster of participants who characterised their use of the platform as somewhere between useful and necessary. Given that few participants positioned the social media platform as a luxury, it is reasonable to conclude that the majority of users recognised that there would be some convenience cost in reducing their usage.

When asked what they saw as the primary benefit of their use of social media platforms, a prominent constellation of participants emphasised the platform's importance for social participation (Figure 8.2). It is important to reflect upon the connection between smart tech and social participation. In many ways the learning benefits of social media platforms are less important for users than they are with other smart devices. Social media platforms may be able to make

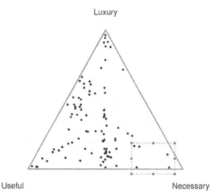

Source: SenseMaker survey.

Figure 8.1   *'How would you describe your use of this social media platform?'*

helpful friend suggestions, flag-up relevant events and news items, or move particularly pertinent posts up your feed. But the primary long-term benefits of social media networks appear to derive from their network effects. Economists describe network effects as the value which users derive from a good or service as a direct result of the number of other people who use the service. While these network effects are connected to the monopolistic status achieved by many social media giants, they may also reflect a form of natural monopoly (would we really want to spread our social connections over multiple plat-forms?). Here, the benefits of social media platforms are less to do with the convenience of having smart input into our actions, and more about the expe-diency of having our social lives coordinated in one place. Reducing our use of social media platforms would inevitably cut people off from these network effects and may reduce their socio-digital capacities and capital. Stopping using such platforms entirely would mean having to extensively rebuild our digital social lives. Of course, rebuilding our digital social networks would not just be inconvenient if we left Big-Tech platforms. The nature of their network effects means that there would be few people available on alternative platforms with whom to rebuild those networks.

According to our survey there is a complex series of factors that may con-tribute to users wanting to reorient their relations with smart technology. When asked to consider the negative impacts of social media platforms on their lives, participants indicated that a loss of focus was a concern (Figure 8.3). Given that the economic model of many smart-tech platforms requires the capturing

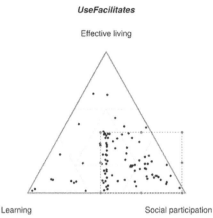

UseFacilitates

Effective living

Learning   Social participation

Source: SenseMaker survey.

Figure 8.2   'Using social media platforms facilitates …'

of as much human attention as possible (to both feed data analysis streams and enable behavioural prompts to be delivered), this concern is to be expected (Kitchin and Fraser, 2020; Zuboff, 2019). Concern over an erosion of focus may not just be about a loss of immediate attention though. Yes, when we become trapped in the endless scroll of our social media feed we are not focusing on other things, people, or tasks. But focus can be disrupted by smart tech in other ways. Our engagement with social media is in part driven by our desire for social connection and the addictive dopamine rewards this can generate. This form of emotional connection with smart tech does not, however, stop once we cease our endless scrolling. The emotional impact of our engagement with smart tech continues as we return to work and family life. This extended form of emotional distraction is often what leads us straight back to social media (to reread that ambiguous post, or check on responses to our provocative Tweet). This process generates an extended field of digital distraction which seeps beyond our direct use of devices. It is in these contexts that smart tech can affect both our direct focus and the quality of attention we can bring to other non-smart-tech things.

Our survey asked participants to reflect on some of the most troubling aspects of smart-tech platforms knowing them well. Figure 8.4 demonstrates that a cluster of participants were equally concerned about the impacts this knowledge could have on their privacy, their ability to control their personal data, and the opportunities it created for others to control their behaviour. When asked to assess motivations for using social media platforms less, there was a cluster of respondents who prioritised personal wellbeing (Figure 8.5).

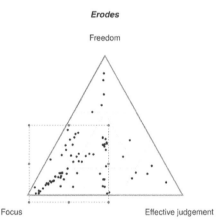

Source: SenseMaker survey.

Figure 8.3   *'Using social media platforms erodes my sense of ...'*

We will explore in greater detail in the following section (on Facebook deleters) what the preservation of personal wellbeing may relate to. There was, however, an additional cluster of respondents who prioritised both personal wellbeing and protection of privacy as motivations for using social media platforms less (Figure 8.6). While some respondents did prioritise time saving as a motivation for reducing their use, it is interesting that, in this survey at least, the time spent on social media was not such a prominent concern. It is, perhaps, wise not to draw any firm conclusions from these results. Nevertheless, it is interesting to position the relative lack of concern over time saving with the previous emphasis that was placed on the loss of focus that social media is associated with (see Figure 8.3). The suggestion here, again, is that the problem with social media is not the absolute amount of time spent on it, but its lingering distractive potential.

When asked how they would characterise their ability to stop using a social media platform, two main clusters of responses emerged. The first suggested that ending use would be easy (see Figure 8.7). Again, this may reflect the relative ease that people find in stopping using one platform, compared to blocking all smart tech in their lives. Or it may reflect the fact that there are just differential social capacities to disengage with social media. But, if Hill's previously discussed experiment is any indication, these responses could also reflect that it is difficult to recognise how fully entwined our lives are with social media until we stop using it. Despite this, there was also a cluster of participants who recognised the complexities that would be involved in removing just one social media platform from their lives (Figure 8.8).

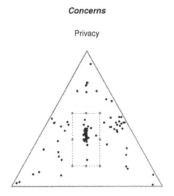

Source: SenseMaker survey.

Figure 8.4   *'What concerns do you have about your platform knowing you well?'*

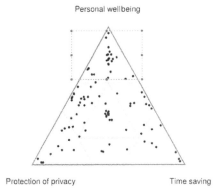

Source: SenseMaker survey.

Figure 8.5   *'What would your primary motivation be to use social media platforms less?' (Cluster 1)*

The SenseMaker survey we completed is helpful not only because of what it shows, but because of how it shows it. The triads above indicate two things about people's relationships with smart technology (in this instance, a social media platform): (1) that their motivations for using it, and their desires for a reorientation in their relationship with it, are complex amalgams of often competing forces; and (2) that people's relationships with smart technology are varied, and while there may be discernible clusters of relations, it is important to recognise the differences that do exist. Our results indicate that

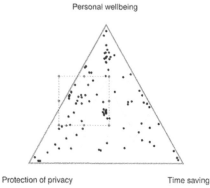

Source: SenseMaker survey.

Figure 8.6   *'What would your primary motivation be to use social media platforms less?' (Cluster 2)*

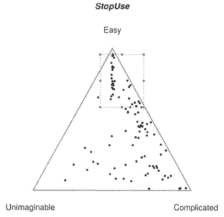

Source: SenseMaker survey.

Figure 8.7   *'How would you characterise your ability to stop using this social media platform?' (Cluster 1)*

the loss of focus associated with social media is an important factor that leads some people to reflect on their relationships with it. This observation would appear to resonate with the personal wellbeing benefits that certain participants associated with using social media less. It is also apparent that different groups of respondents view their ability to stop using social media platforms very differently. Survey participants did, however, give consistent prominence to the role of social media in facilitating social participation. Given the powerful net-

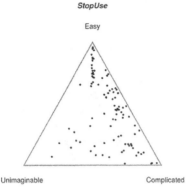

Source: SenseMaker survey.

Figure 8.8   *'How would you characterise your ability to stop using this social media platform?' (Cluster 2)*

working effects of such platforms, it would seem that any form of smart-tech reorientation will come with social costs to many.

Building on these insights, the remaining section of this chapter considers the motivations of one particular group who have attempted to reorient their relationship with smart technology.

THE CASE OF FACEBOOK DELETERS

This final section of the chapter reflects on a study of so-called Facebook deleters. This study was carried out during 2019 and involved in-depth interviews with people who had deleted, or in some way stepped away from, their Facebook account. This study was conducted to get a better sense of the varied personal motivations that informed people's decisions to leave social media platforms and to reorient their relations with smart technology. While this study thus builds on the insights of the SenseMaker survey reported in the previous section, it also takes an additional analytical step. In addition to considering people's motivations for leaving social media, the study also explored the consequences associated with this action. There are obviously clear lines of connection between this study and Kashmir Hill's Big-Tech block. By conducting in-depth interviews with a number (ten) of Facebook deleters, however, the study was able to consider a more varied set of motivations for, and consequences of, smart-tech blocking.

The #DeleteFacebook Movement

There is a tendency to think of the Delete Facebook Movement as a relatively recent phenomenon. However, since its launch in February 2004, there has been a series of moments when the social desire to delete Facebook appears to have been strong. Over the last decade there has been a series of prominent spikes in interest about Facebook deletion. These spikes correspond with various political, economic, and social controversies associated with the company. In 2012, for example, there were various bad news stories in circulation about Facebook, including those pertaining to its tax-avoidance activities, research connecting the platform to problems of stress and addiction, and revelations that there were 83 million fake accounts on the platform. The 2018 spike in Facebook deletion interest is, however, of most relevance to research recounted here. In March 2018 news broke of an investigation by *The Observer* newspaper into activities of a company named Cambridge Analytica (Cadwalladr and Graham-Harrison, 2018). The investigation revealed that the data analytics company had been able to harvest somewhere in the region of 50 million Facebook user profiles (at the time about one-third of all active North American Facebook users; Cadwalladr and Graham-Harrison, 2018). Facebook user data was then used to build profiles for targeted political advertising ahead of the 2016 US presidential election. Crucially, although Facebook became aware of the unusual scale of the data harvesting undertaken by Cambridge Analytica, it did not alert users (Cadwalladr and Graham-Harrison, 2018). These accusations were revealed at the same time that news was emerging of the use of Russian exploitation of Facebook to try to influence the US presidential election.

The Cambridge Analytica scandal energised the #DeleteFacebook Movement, which gained momentum during 2018. In exploring the experience of Facebook deleters in 2019, the working hypothesis of our research was that it was scandals such as the one surrounding Cambridge Analytica that were driving people from the platform. We surmised that Facebook deletion was a product of smart tech being used to undermine privacy and manipulate political opinions and behaviours. There is circumstantial evidence to support this hypothesis. Research carried out shortly after the scandal (in May–June 2018) by the Pew Research Center revealed some immediate user responses. This research revealed that 54 per cent of surveyed users had adjusted their privacy settings during the previous 12 months (a period that in part predated the Cambridge Analytica revelations); 42 per cent had taken some sort of break from the platform (for several weeks or more); and 25 per cent of those surveyed had deleted the app from their phone (Perrin, 2018).[2] We also know that Facebook's market capitalisation dropped by over $100 billion in July of 2018 (a product of slowing user growth; Neate, 2018). Despite these trends, our research uncovered a much more complex set of motivations for leaving Facebook that speak directly to this volume's concerns with the emerging relations and tensions between humans and smart tech.

Why Do People Leave Facebook?

To understand why people leave or delete Facebook it is necessary to interrogate what leaving actually means. It turns out that leaving and/or deleting Facebook are not binary actions but involve a range of different digital reorientation strategies. Some of the people we spoke to did stop using Facebook entirely and actually deleted their account and associated records. Of those we have spoken to, however, this is a relatively rare form of action. Even in the case of account deletion, it was clear that leaving Facebook was often a gradual process. As one interviewee reflected:

> What was the trigger? I mean there was the campaign #DeleteFacebook, that was the point at which I deleted my account, but mentally I had disengaged in the two years before, I think I had probably used it a handful of times [since the Brexit referendum and 2016 US presidential election] ... If you want a distinct trigger, the trigger was, this was the week that Zuckerberg had been summoned to Parliament, and I think that was [when] the Delete Facebook campaign really took off at that point. It seemed to me that this was a good moment to add my voice. (Male Facebook deleter interview, 2019)

In this instance, account deletion was a kind of political capstone to a much longer two-year period of gradual disengagement from the platform. Here, the actual act of deletion appeared to serve as a discrete political rather than a social goal. Ambiguous forms of disengagement, which defy binary on–off or blocked/unblocked logics, were a common theme within the interviews we conducted.

Only one of the interviewees we spoke to conformed to the pattern of Facebook deletion that we anticipated at the start of the research. This interviewee described how on 17 March 2018, directly following the breaking news of the Cambridge Analytica scandal, they went upstairs and deleted their Facebook account, and ensured that the patches of paper that were covering their computer camera were in place. But even in this seemingly straightfor-

---

[2]   These survey results were taken from the Pew Research Center's American Trends Panel. This particular survey involved 4594 respondents.

ward account of deletion there is a more complex backstory. This Facebook user had actually deactivated their Facebook account in 2017. They returned to the platform soon after deactivation, only to be finally persuaded to leave following Cambridge Analytica.

Most of those we spoke to described the act of leaving Facebook as something that stopped short of full account deletion. Many just stopped posting things or checking their feed. Some people took quite drastic steps to stop their compulsion to check Facebook, which still stopped short of deletion. One interviewee described getting her friend to change the password on her account to prevent her from logging on. Others simply deactivated their accounts but stopped short of deletion just in case they decided to return to the platform in the future. The following reflection from one Facebook user is instructive in this case:

> I planned to leave Facebook altogether. So, a couple of months ago I backed up all my Facebook data, which I think they are entitled to keep copies of ... so I downloaded it all, then I thought ummm, in the organising way where it is useful to me, maybe I should stay on. So, what I have done is posted a blank; you know, I don't post anything on my page anymore, I don't really look at Facebook except to follow events I am coordinating or involved with. Of course, I do then see that someone has messaged me. (Male Facebook deleter interview, 2019)

Many self-identified Facebook deleters we spoke to talked about 'leaving' Facebook on these kinds of terms. They wanted to leave Facebook, they certainly wanted to use it less, but in the act of deactivation they stopped short of full deletion.

There are of course many explanations for this inability to completely delete Facebook. In the quote above we can see how Facebook had become central to the coordination of social events (in this instance, choir practice) that this user was involved in. Here the network effects of Facebook are a powerful force keeping people connected (it is much easier to organise events on a platform that everyone is on). When blocking Facebook, Kashmir Hill reflected on the following challenges of stepping away from the platform: 'Like many people, I feel invested in Facebook: I've been building my profile since 2007. I have party and vacation photos galore there and over 1000 connections, including dear friends, acquaintances, colleagues, loved ones, and quite a few randos whom I added for reasons that I no longer remember' (Hill, 2019f). Here, Hill draws attention to the role of Facebook as a kind of accidental archive of our lives. As a form of transactive memory, deleting Facebook can be seen as a loss of a part of ourselves (part that we may not even be consciously aware of). Writing in *The New Statesman*, Pippa Bailey (2020) observed that 'Deleting my [Facebook] profile felt like a loss because it kept the intimate details of a life I no longer remember'.

Various forms of sunk-cost biases and loss aversion serve to make a full Facebook deletion psychologically challenging. At the very least they appear to make temporary deactivation more appealing than deletion. It is clear that this deactivation stage is being exploited by Facebook as a way of wooing potential deleters back into active use of the platform. As one interview reflected,

> There were already, certainly at that point [following the Cambridge Analytica scandal in 2018], people who were trying to leave, were finding messages at the end of the leaving process, you know, 'These are your friends, they are sad to see you go'. And then actually the mechanics of leaving were also, it just amazed me that it was even legal, you know? You actually weren't, you hadn't left, you were merely deactivated, and at any point if you were to log on to Facebook or something associated with Facebook, you were counted as activated without being asked. (Male Facebook deleter interview, 2019)

The evident use of social influence and presumed reactivation consent are, of course, classic tools of behavioural manipulation (see Chapter 4). The fact that these tactics were being used by Facebook to reactivate disenchanted users illustrates that smart technology can still wield significant influence even when you have placed one foot outside of its digital door.

So, to be clear, when considering the act of leaving Facebook, it is best to think of it as a varied zone of digital reorientation. This zone includes full and unequivocable account deletion, temporary account deactivation, more flexible Facebook holidays, and simply leaving an account idle. Although those involved in each of these actions identify with Facebook deletion, they are perhaps best thought of collectively as Facebook leavers.

*

The reasons for leaving Facebook appear to be as diverse as the ways in which people leave. A key insight of our study was that concerns over privacy breaches and Big-Tech scandals tended (with one or two exceptions) to not be the primary motivation for leaving Facebook. Some Facebook leavers were simply not aware of major privacy scandals. As one participant observed, 'I'll be honest, you sent that [message mentioning Cambridge Analytica] and I had no idea of what [that] was, so I had to go away and Google it' (Female Facebook leaver interview, 2019). Other leavers reflected on the fact that these scandals had served to emphasise what they already knew: 'I think it solidified my decision to leave ... although I don't know enough about it but I kind of ... my initial reaction was, "well yea, of course they sell our data". I was a bit, like, people don't know this?!' (Female Facebook leaver interview, 2019).

When privacy was a motivation for leaving, it was often the lack of trust people had with their own expanded friend network on Facebook than with the selling of more generic forms of personal data to third parties. Here is one related conversation with a Facebook leaver:

> Interviewer: How many friends had you got to?
> Leaver: Over a thousand. And the majority of them I had known, but that could be meeting them once or seeing them with a friend of a friend. They weren't the people I would spend time with. Yes, over a thousand. And that was actually the start. I looked at those people and I thought, if I put a picture of my son up, do I want them to see him? Would I go to them personally and say 'this is my son?' And the answer was no! So, I started culling friends and I got down to 500, and I thought if I am doing this because I am scared about my privacy, yet those pictures are still accessible for people, then I may as well not be on it. And that was the point at which I left. (Female Facebook leaver interview, 2019)

There is something sociologically significant about our inability to trust the size of the social networks that are now becoming associated with social media platforms like Facebook. In the 1990s, the British anthropologist Robin Dunbar (1992) suggested that humans generally had the (cogitative) capacity to maintain about 150 stable social relations. While this number, and its scientific basis, is disputed, it does find echoes in corporate organisational patterns, military units, and even in the size of villages. Dunbar's number has significance beyond the issue of social trust. It is also indicative of the levels at which the forms of social obligation required to maintain meaningful relations can be sustained.

In this context, it is significant that one of the reasons given for leaving Facebook was the form of obligations it generated:

> Quite early on I did get the sense that the sort of mutual obligations that arose became onerous. I wasn't that interested in cats. It always came up a lot as we are a family of cat lovers. I am not that interested in the children either … there is a whole psychological system that arises that hasn't before. And I think this is the same with, you know, the two-tick system. There may only be two or three elements in play, whether it is: knowing when a message has arrived; knowing it has been seen: has it been commented on? But that is enough actually for people to build up a sense of resentment, or isolation. (Male Facebook leaver interview, 2019)

The social etiquette associated with responding to communications (and forms of gift-giving) has evolved over many millennia. Indeed, the acts of responding to communication and nurturing friendships have important evolutionary advantages. In the era of Facebook, and the expanded social networks that related platforms enable, a tension tends to emerge. This is a tension between our continued desire to give and receive forms of social recognition, and our ability to provide such recognition and affirmation to others. Our interviewees described the digital demands associated with maintaining virtual networks. They also described the psychological costs of not seeing their digital sociability reciprocated. In one context these concerns were expressed in relation to people simply not responding to general posts on Facebook:

> I left Facebook in the end for purely personal reasons, because another thing, it becomes an echo chamber, so sharing *Guardian* posts, or *New Statesman* posts, or some political view in a quick, glib way, and then somehow being addicted to seeing how many likes you get, and then getting no likes, and then somehow feeling bad about it and yourself, and wasting a lot of time checking email every ten minutes and Facebook every 20 minutes. So, you think, where are all the people who are hanging on my every word? And then I think, well I am not hanging on anybody else's word … so why am I expecting to be some kind of viral Facebook guru with people very interested in me? *But it does leave you with that kind of lonely, isolated feeling, more than the connected feeling. It leaves you feeling more disconnected than connected I think.* (Male Facebook leaver interview, 2019, emphasis added)

The feelings of being digitally overlooked and disconnected described above are a product of the social labours that Facebook demands of its users. As digital social networks grow, users are likely to tire of maintaining social connections, and in turn are likely to feel the forms of social isolation that occur when social links are not upheld.

But the forms of social exclusion manifest on Facebook also take more deliberate and pernicious forms. One respondent described how Facebook had, over a relatively long period of time, served to consistently remind her of her exclusion from certain social networks which she had once been a part of. The sharing of images of social gatherings (including trips away and weddings), of which everyone in the social group appeared to have been invited but our respondent, were a cause of anxiety and sadness to her. It would ultimately be her primary motivation to leave Facebook—she felt it was better for her to remove herself from her entire Facebook network then be algorithmically reminded of the real-world networks she had lost. The exclusion of individuals from social groups is as old as society itself. What Facebook facilitates is an unusual ability to make these forms of exclusion and their precise nature (in this instance, the exclusion of just one person from an entire group) transparent. While conventional social norms of course would generally obscure these forms of social rejection, Facebook is socially unintelligent when it is informing Facebook friends of their downgraded friendship status in the real world. This example reveals how smart tech can reveal certain social truths that are deliberately obscured within real-world relations. Such digital oversights are likely to lead to unnecessary forms of social division that are likely to be in no one's best interests.

Many respondents discussed the broader forms of emotional labour that are associated with using Facebook. This emotional labour went beyond reciprocity/non-reciprocity to include heightened forms of social comparison. Social comparison is, of course, an inevitable part of life, as we strive to assess our relative standing within social networks and wider society. But on Facebook (as with many other smart-tech platforms) the capacity for social comparison is greatly enhanced. One respondent reflected on the psychological costs of the social comparisons that Facebook facilitates:

> I know I was looking at it [Facebook] too much. I was going through a period of anxiety, and I think this is very common—when you are feeling anxious, your self-esteem tends to be quite low and then it is that comparing yourself with every-body ... and even though I am an intelligent person and I know that what I see on social media isn't real, it feeds that part of me. It's like everyone else is made, is more successful, is happier, all that stuff ... And I got to a point where I thought, nah, it's not good for me. And I had had a period, I think over Christmas, where I stopped using Facebook and I did feel a lot better, so I thought, I am going to try it again, and I haven't been back on it. (Female Facebook leaver interview, 2019)

A related reason that several informants gave for leaving Facebook was the issue of oversharing. It appears that as people become more used to using social media platforms, and their addictive qualities become established, that certain users share increasingly intimate information about their everyday lives. While these forms of social sharing may be acceptable in smaller groups of friends and family, many found that they crossed conventional boundaries of intimacy on Facebook. Oversharing seemed to produce two main responses. First was a form of embarrassment that users knew too much about relatively distant Facebook friends. Second, was a sense that Facebook increasingly offered users content that was of less interest and utility to them (except in fairly voyeuristic terms). As with many of the reasons that people gave for leaving Facebook, this issue is connected to a perpetual drive to grow the social network (Hao, 2021). Growth appears to be crucial to smart technology because it both enhances its predictive intelligence and supports the accrual of increased profits. At the same time, this growth appears to be reaching certain social limits pertaining particularly to the maintenance of social networks and intimacy. While this may not threaten the economic model of Facebook right now, it could well do in the future. Such developments appear to be opening up market opportunities for new smart apps.[3]

---

[3]   Following our sharing of some of the insights presented in this chapter (Whitehead, 2020), we were contacted by one app developer who is trying to design a new social media platform that recognises limits to the size of our digital social network and aligns itself with established sociological conventions.

The Consequences of Facebook Deletion

In addition to asking interviewees how and why they left Facebook, we were also interested in the impacts leaving and/or deletion had on their lives. Common responses we received emphasised the wellbeing benefits of being off the platform. These wellbeing benefits related to a relieving of the pressures and self-assessments that are part and parcel of the forms of social comparison which social media prompts. They were also related to the lightening of the cognitive and emotional loads associated with the maintenance of extended social networks. Interestingly, in Hill's previously discussed tech block, she noticed that when the need to monitor and maintain her digital network was removed, she felt much more connected to her family (Hill, 2019f). Specifically, she noticed a discernable increase in her ability to devote more attention to her young daughter. It appears that stepping off social media platforms can generate benefits that derive both from a disengagement from digital responsibility and a reengagement with more proximate real-world interactions. Such observations would appear to support the purported wellbeing benefits associated with the deceleration of our lives promoted within the Slow Computing Movement. We should not be surprised that Facebook leavers report such wellbeing benefits. Numerous studies have highlighted the highly addictive qualities of the platform and how leaving it is similar to giving up smoking or alcohol.[4] It is important to note that Facebook is very much aware of the negative impacts that the platform has on users' (particularly more vulnerable users') mental health (Hao, 2021). It has even considered the steps it could take to moderate the content of those who are algorithmically identified as showing signs of anxiety and depression (Hao, 2021). It appears that Facebook executives are only willing to countenance such measures when they do not affect user engagement with the platform. When, as our study has shown, the wellbeing problems of Facebook derive from routine engagement with the platform, and not just distressing content encountered there, it appears unlikely that smart technology will be able to solve this problem on its own (Hao, 2021).

---

[4] One of the earliest studies to highlight the addictive qualities of social media was the 'World Unplugged' study conducted by the International Center for Media & the Public Agenda (ICMPA) and the Salzburg Academy on Media & Global Change in 2011. This study asked approximately 1000 international students to abstain from using social media for 24 hours. The study revealed how difficult it was for these students to disconnect, and uncovered connections between social media use and anxiety and depression. Interestingly, the study described mobile phones as this 'generation's Swiss Army knife AND its security blanket' (see: https://theworldunplugged.wordpress.com).

Other respondents reflected on the time-saving benefits associated with leaving Facebook. Leaving Facebook was associated with the freeing of more time during the day for work and domestic tasks as the compulsive elements and procrastinating temptations of Facebook were allayed. This, of course, reflects the forms of time sovereignty that the Slow Computing Movement speaks about. Interestingly, and related to this observation, it was also noted that leaving Facebook resulted in people being drawn in to digital activities less frequently. In this context, one respondent reflected on the fact that he was now reading more books (which helped him to think) and far less online material. In these terms, it is important to remember that social media platforms, as with many other smart-tech devices, are gateways to all manner of digital activities. If you close the gateway, the pull of digital life is partially ameliorated.

Respondents to our study, however, also reported inevitable negative consequences to leaving Facebook. Prime among these was the issue of social disconnection. Hill describes how, during her Facebook block, she missed out on the announcement of the birth of a child to her friend—the friend had assumed that posting the announcement on Facebook meant that everyone she knew knew (Hill, 2019g)! Hill goes on to reflect, 'And I realize I don't really know what people are up to. My friends now largely expect that I'll see their broadcasts on various social networks, which means they don't tell me things individually anymore, unless I see them in person' (Hill, 2019g). The processes of social disconnection associated with leaving Facebook are thus not merely a product of not being included in digital communication. They also stem from the fact that social media makes people less inclined to share information in traditional ways. It is in this context that we are reminded, once again, that changing our relations with smart technology is a collective-action problem.

The problems of social disconnection appear to be one of the main reasons that people choose to leave but not delete Facebook. One of our respondents described how she needed to stay within Facebook's digital ecosystem so that she could continue to use Messenger (as that was how she stayed in touch with her mother). Furthermore, leaving Facebook often involved moving to other direct messaging services such as WhatsApp. While moves to a more restricted forms of digital communication and data sharing enable people to assert new forms of time and data sovereignty, it does not mean that they are free of the processes of data extraction.

CONCLUSION

In this chapter we have explored various strategies that people are using to dumb down and de-convenience their lives as they reorient their relations with smart technology. These strategies range from the blocking and deleting of smart tech to the more complex technological reorientations associated with movements like slow computing. What appears to be clear is that the human capacity to reorient our relations with smart tech is already heavily circumscribed by our dependencies on it. Furthermore, reorienting human–smart-tech relations is already a collective-action issue and not merely one of individual consumer choice.

It is tempting to connect changes in our individual and collective relations with smart tech with controversies such as the Cambridge Analytica scandal. Our research, however, demonstrates that dumbing down is more commonly a product of the social and psychological costs associated with using smart-tech platforms on a regular basis. These social and psychological problems appear to be products of the expansionist logics of Big-Tech. While the expansion of smart-tech networks facilitates the growth of digital learning and surveillance capitalist revenues, it generates problems of trust, social obligation, and anxiety for users. How smart-tech systems detect and react to the human problems that its modes of operation generate could be critical to the future trajectory of the smart-tech society. If smart tech cannot support long-established human needs for trusted and manageable social networks—in other words, become more human—disconnection may become the norm and not the exception in the future.

9. Conclusion

TRANSITION DEMANDS AND HUMAN RESPONSIBILITY IN THE INTELLIGENT-TECH SOCIETY

As we neared the end of writing this book our minds naturally turned to what our conclusions might be. At the same time, the Law Commission for England and Wales (with the Scottish Law Commission) made proposals for the establishment of an Automated Vehicles Act in response to the emergence of self-driving vehicles (Law Commission, 2022). The proposed Automated Vehicles Act seeks to make the manufacturers of automated vehicles legally responsible for accidents that occur when a car was in self-driving mode. In many ways, the Law Commission Report embodies a response to one aspect of the legal challenges associated with questions of human responsibility in the smart-tech society. A common refrain throughout this volume has been that the unerring rise of smart technologies, such as the self-driving car, has emerged because of a belief in the significant social and economic benefits such technologies bring (the self-driving car has been associated with myriad advantages, including safety, convenience, and environmental protections). These developments have, however, brought with them new implications for how we understand human nature, autonomy, and responsibility. In terms of self-driving vehicles, the question then arises as to how humans can be held responsible for what their car does when their attention is on more (or less) pressing matters.

What interests us most about the Law Commission Report is the litigious attention it brings to the precise parameters of human responsibility in self-driving cars and the implications of so-called transition demands. The Law Commission argues that humans should not be held criminally responsible for offences caused by self-driving vehicles when the vehicle is in self-driving mode. More controversially, however, the Commission also assert that in certain circumstances drivers should be exempt from prosecution even when they have assumed control of the car. The Law Commission use the following example to demonstrate their reasoning:

> While in self-driving mode, an automated vehicle turns into a one-way street in the wrong direction. The user-in-charge takes over but is unable to avoid a collision. Alternatively, no collision takes place, but in the moment the user-in-charge takes over, they are driving in the wrong direction and may be guilty of an offence on that basis. We recommend that the driver should have a specific defence to any driving offence committed in the period immediately following a handover. (Law Commission, 2022: 18)

While intuitively sensible, this legal recommendation raises broader existential questions for the smart-tech society. If we are to collectively adopt the Law Commission's principle here, it is clear that issues of responsibility in the human–smart-tech interface are not binary. There may well be times that even when in control, humans may still be dealing with the consequences of smart-tech choices and thus not be responsible for what then happens.

The Law Commission also raise the question of so-called transition demands. Transition demands are used within smart/self-driving vehicle operating systems to request that the human assumes full responsibility for the vehicle (Law Commission, 2022: 18). According to the Report, the simple request for a transition of responsibility does not mean that responsibility has moved from the car to the human. The Commission argues that there must be a transition period which enables a human to be able to realistically reassume control of the vehicle. For us, the very notion of transition demands reflects a significant perspective on the human–smart-tech interface. It embodies a moment when smart technology 'demands' human engagement. Of course, a transition demand (and any associated transition period) works both ways. What about when a human requests smart technology to take control of certain things? In such circumstances, precisely when does the responsibility pass back from humans to machines, and what of the lingering consequences of human actions on the smart technology's ability to act?

For us, these legal discussions about self-driving cars are emblematic of deeper issues within the smart-tech society. They speak to the complex, circumstantial, and highly contextual nature of human–smart-tech relations. They also reveal the limitations of simplistic and binary accounts of smart technology and its human implications. Smart technology cannot be thought of as merely good or bad, on or off, or even indifferent when it comes to human life. If nothing else, the accounts of smart technology that run through this volume reveal that the smart-tech society is not about the technological domination or liberation of humans, it is about myriad relations of learning, control, and communication. These relations have deep implications for established norms of justice, privacy, autonomy, representation, and dignity, inter alia. But these are relations that we believe humans still have the capacity to collectively impose transition demands of their own upon.

*

The notion of transition demands provides a helpful context for thinking strategically about and orienting our interventions within the smart-tech society. In this context we are primarily interested in human demands for a transition of control and not one that emanates from a flummoxed smart-tech system. A transition demand suggests a nuanced and flexible way of conceiving of the human–smart-tech interface. A transition demand is not the same as a kill switch through which humans can ultimately seize back control from smart technology. Transition demands recognise the mutual interdependencies of humans and smart tech, and emphasise the potential benefits of allowing more or fewer human inputs into the technological realm in different situations. We have seen throughout this volume that the augmented capacities of human and smart tech can often produce the best outcomes. This does not necessarily mean that human–smart-tech relations have to be defined by a constant need to balance out human and smart-tech input. In some instances it is clear that greater human input into smart cars, homes, and workplaces may be beneficial. At others, minimising human input may benefit the technological process and human wellbeing. What the idea of a transition demand does draw attention to is the importance of human vigilance towards smart technology. Although we may want to make our lives more convenient and optimise our decision-making through the deployment of smart technology, this does not mean we have to be sidelined in discussions about the strategic form of the smart-tech world we live in. It is thus, perhaps, helpful to think of transition demands in both local and more macro contexts. At the local level, practical vigilance is necessary to ensure that our use of smart tech does not deprive us of privacy and autonomy. In a more strategic context, it appears that we may be in something of a transition window in the evolution of the smart-tech society. During this window we continue to have an ability to shape and resist digital technology's impacts on our ways of life and political systems. Invoking a transition demand in this context is more than merely requiring meaningful human input into smart-tech systems; it is about shaping the smart-tech transition itself. This appears to be something that we will not be able to do individually. It will require the collective institutions of our communities, governments, and legal systems. It will also require a powerful assertion of a collective transition demand that will serve to protect our individual rights to such demands in the future.

Maintaining vigilant forms of engagement with smart technology at both individual and collective levels will clearly not be easy. As the Slow Computing Movement has demonstrated, maintaining our right to demand transitions will often be synonymous with a deliberate inconveniencing of our lives. Given our apparent dislike of inconvenience, this should perhaps be a concern. But, throughout this volume we have seen that the smart-tech society provides us with ample prompts to be critical and vigilant towards its operations. In Chapter 4 we considered a predictive paradox whereby the more that smart technology gets to know us, the more suspicious of it we become. In Chapter 7 we observed how older residents in Singapore's public housing areas were cautious about adopting smart technology in their homes because it reflected an unwanted change and they felt their lives were already good enough. This idea of things already being good enough without smart-tech input is a significant insight. Throughout many of the chapters in this volume we have seen how the use of smart technology only brings relatively marginal gains to the end user (with the most significant gains being presented to surveillance capitalists, or the data elites in government and the workplace). The lack of significant long-term benefits from smart technology for everyday users, alongside the unwelcome changes it often brings to our lives, could offer valuable experiential prompts to vigilant engagement.

*

A recurring theme throughout this volume has been the question of learning. In her dissection of surveillance capitalist systems, Zuboff (2019) reveals a significant division of learning being produced by contemporary smart tech. This division of learning results in data elites and surveillance capitalists being able to abstract far more meaningful forms of learning from smart-tech interfaces than everyday users can. But there is clearly more at stake than aggregate divisions of learning within the smart-tech world. In this volume we have uncovered different styles of learning that are evident in our smart-tech societies. Knox et al. (2020) identify a particular branch of instrumentalised learning in smart educational contexts that is, perhaps, synonymous with the wider pedagogic practices of the smart-tech society. This is a model of pedagogy that deploys machine learning to better understand how humans learn (Knox et al., 2020: 35). Knox et al. (2020) argue that this is a form of smart pedagogy that does not prioritise the rational intents and motivations of human learners, but what best promotes measurable learning behaviours. This behavioural vision of the learner is, of course, synonymous with many of the behaviouralist assumptions that are associated with smart technology (see Chapter 4, this volume). It involves a vision of the human learner that emphasises human irrationality and emotion over intent and deliberation. This model of human learning can be used to hypernudge us beyond our learning biases to unlock new educational opportunities. But as a pedagogic vision, it also carries with it dangers.

In Chapter 4 we speculated that the behaviouralist orientation of many smart-tech platforms may not just be responding to the cognitive flaws of humans but reenforcing them. In this context, it is clear that if smart tech promotes a behaviouralist approach to societal learning this could see a dis-connection emerge between human learning and humans being able to actively determine what they need to learn. We explored some of the implications of instrumentalised forms of human learning in Chapter 6. There we explored how the emergence of smart workplaces is often associated with the compart-mentalisation of knowledge, and more limited capacities of humans to under-stand labour processes in their entirety. While knowing just what you need to know may optimise some aspects of workplace efficiency, it tends to diminish the input that humans can have into labour processes and to undermine estab-lished notions of professionalism (see Pasquale, 2020). But what if being a good citizen in the smart-tech society involves more than just completing your citizenly duties? What if it involves an active determination of what it is to be a good citizen (see de Dijn, 2020)?

In our discussion of smart body technology and self-quantification (Chapter 5), we considered the limitations of systems of smart-tech learning that are premised only on what can be quantifiably measured on and in the body, and that pay no attention to what the human subject may also be able to reveal. In Chapter 5 we also saw that smart-tech learning does not have to be narrow, instrumental, and disempowering to humans. The combination of digital and human perspectives on the body can provide an augmented perspective on health and the body which far exceeds the forms of knowledge that were available before the advent of smart technology. One of the emerging chal-lenges of the smart-tech society, then, is to think about the models of learning it should be based upon. At present, it seems that the 'smart' in 'smart tech' is synonymous with a form of learning that prioritises the operational needs of digital technology and seeks to minimise cognitive engagements from humans. But what if the smart-tech society could aspire to higher forms of augmented intelligence? It is not difficult to envisage a form of intelligent-tech society in which there was not a zero-sum game in operation between human and technological learning. Indeed, it is likely that digital tech could not only enhance human knowledge but also help us achieve higher levels of learning and meaning making. In this intelligent-tech world, digital learning would not only be used to help us control our cognitive weaknesses and behavioural impulses, but also transform our understandings of the world and our places in it (see Kegan, 1982). If nothing else, it seems much more likely that we are going to invoke transition demands of smart technology if that very same technology is helping us to think more critically about the different worlds it could help us to build.

*

This book has focused on an interface (in truth, a multitudinous set of inter-faces) that connects humans with smart technologies. As we come to the end of this volume, we are aware of the aspects of this interface we have not been able to cover. We are conscious, for example, of not exploring the highly gendered and racialised aspects of this interface and how smart tech is experi-enced very differently by different people (see Strengers and Kennedy, 2020). But where we hope we have been successful is in revealing the complexities and radical uncertainties that characterise this interface. The interface between humans and smart technology is clearly a dialectically adaptive one, wherein smart tech is transformed by human input (and resistance), and human life and wider society are altered by our individual and collective engagements with digital technology. Although there is clearly adaptative capacity left in our smart-tech society, we would argue that it is best to think of this capacity less as an open road and more as a closing window. If left unchecked, the evolution of our smart-tech societies will undoubtedly slow and solidify, and our already circumscribed capacity to shape its contours will undoubtedly diminish. This is precisely why vigilance towards and engagement with smart-tech systems should not be seen as just a long-term goal.

In his 2012 volume *Missing Out: In Praise of the Unlived Life* the psycho-analyst Adam Phillips argues that humans are defined by two lives: the one we live and the one we think we might like to live. In many ways this gap between our actual life and the one of fantasies is drawn into sharp focus by smart tech-nology. Smart technology provides numerous windows on an alternative life: whether through our social media feeds or the promises of labour-free lives of opportunity and convenience. Interestingly, Phillips claims that not achieving our dreamed-of alternative life may be important for our wellbeing and living a full life. The fantasy is what drives us; its fulfilment would be a potentially devasting disappointment. And so it may well be with the smart-tech world. The promise of our smart-tech lives—of a more connected, more fulfilling, more effective existence—is perhaps what is most important. It seems very unlikely that smart tech will ever be able to deliver the life of our dreams. Even if it could, perhaps it would not be in our own best interests. It does seem likely, however, that if left to its own devices, Big-Tech could produce a world that is disempowering and disappointing. This is why critical engagement with emerging smart-tech systems is so important. Such engagement could energise our collective efforts to forge a better future. Furthermore, if we don't question and shape our collective transition to a smart-tech world now, we may find it much more difficult to do so in the not-too-distant future. So, our message is simple: Engage now!

References

Akerlof, G. and Shiller, R. (2009) *Animal Spirits: How Human Psychology Drives the Economy and Why it Matters for Global Capitalism*. Princeton, NJ: Princeton University.

Alston, P. and van Veen, C. (2019) 'How Britain's welfare state has been taken over by shadowy tech consultants', *The Guardian*, 27 June. https://www.theguardian.com/commentisfree/2019/jun/27/britain-welfare-state-shadowy-tech-consultants-universal-credit (Accessed 7 April 2022).

Amoore, L. (2017) 'What does it mean to govern with algorithms', Antipode Intervention Symposium. https://antipodeonline.org/wp-content/uploads/2017/05/2-louise-amoore.pdf (Accessed 9 April 2022).

Amoore, L. (2020) *Cloud Ethics: Algorithms and the Attributes of Ourselves and Others*. Durham, NC: Duke University Press.

Anderson, C. (2000) 'The end of theory: The data deluge makes the scientific method obsolete', *Wired*, 23 June. https://www.wired.com/2008/06/pb-theory/ (Accessed 7 April 2022).

Bailey, P. (2020) 'Why I finally deleted my Facebook account after watching Netflix's *The Social Dilemma*', *The New Statesman*, 21 October. https://www.newstatesman.com/science-tech/2020/10/why-i-deleted-facebook-account-the-social-dilemma-netflix-documentary (Accessed 7 April 2022)

Ball, J. (2020) *The System: Who Owns the Internet, and How it Owns Us*. London: Bloomsbury Publishing.

Ball, J. (2021) 'Automatons of the world unite! How the Party of Workers began talking about the abolition of work', *The New Statesman*, 17 September. https://www.newstatesman.com/spotlight/2021/09/automatons-of-the-world-unite (Accessed 12 August 2022).

Bank of England (2021) *Will a Robot Takeover my Job?* https://www.bankofengland.co.uk/knowledgebank/will-a-robot-takeover-my-job (Accessed 7 April 2022).

Barbrook, R. and Cameron, A. (1996) 'The Californian ideology', *Science as Culture* 6: 44–72.

Barkan, J. (2013) *Corporate Sovereignty: Law and Government under Capitalism*. London: University of Minnesota Press.

Barlow, J.P. (1996) 'A declaration of the independence of cyber space', *The Electronic Frontier Foundation*. https://www.eff.org/cyberspace-independence (Accessed 3 March 2022).

Bartlett, J.P. (2018) *The People vs Tech: How the Internet is Killing Democracy (and How We Save It)*. London: Ebury Press.

Bastani, A. (2019) *Fully Automated Luxury Communism – A Manifesto*. London: Verso.

BBC (British Broadcasting Corporation) (2020) 'Amazon faces backlash over Covid-19 safety measures', 17 June. https://www.bbc.co.uk/news/technology-53079624 (Accessed 7 April 2022).

Beer, D. (2019) *The Data Gaze: Capitalism, Power and Perception*. London: SAGE.

Berreby, D. (2021a) 'Behind every good robot there's a human', *Robots for the Rest of Us*, 15 July. https://robots4therestofus.substack.com/p/behind-every-good-robot-theres-a?s=r (Accessed 7 April 2022).

Berreby, D. (2021b) 'Is it ever OK to bash a robot?', *Robots for the Rest of Us*, 14 February. https://robots4therestofus.substack.com/p/is-it-ever-ok-to-bash-a-robot (Accessed 7 April 2022).

Bolluyt, J. (2014) 'Here's why Facebook is ending election day experiments', *Tech Cheatsheet*. https://www.cheatsheet.com/technology/heres-why-facebook-is-ending-election-day-experiments.html/ (Accessed 18 July 2019).

Bond, R.M., Farris, C.J., Jones, J.J., Kramer, A.D.I., Marlow, C., Settle, J.E. and Fowler, J.H. (2012) 'A 61-million-person experiment in social influence and political mobilization', *Nature* 489(September): 295–8.

Bray, T. (2020) 'Bye, Amazon', 29 April. https://www.tbray.org/ongoing/When/202x/2020/04/29/Leaving-Amazon (Accessed 28 March 2022).

Buettner, D. (2012) *Blue Zones: 9 Lessons for Living Longer From the People Who've Lived the Longest*. Washington, D.C.: National Geographic Society.

Bukatman, S. (1997) *Blade Runner*. London: British Film Institute.

Burgess, M. (2018) 'What is the Internet of Things – *Wired* explains', *Wired*, 6 February. https://www.wired.co.uk/article/internet-of-things-what-is-explained-iot#:~:text=IoT%20allows%20devices%20on%20closed,a%20much%20more%20connected%20world.%22 (Accessed 12 August 2022).

Butler, P. and Walker, P. (2016) 'Universal Credit falls five years behind schedule', *The Guardian*, 20 July. https://www.theguardian.com/society/2016/jul/20/universal-credit-five-year-delay-2022-damian-green (Accessed 7 April 2022).

Cadwalladr, C. and Graham-Harrison, E. (2018) 'Revealed: 50 million Facebook profiles harvested for Cambridge Analytica in major data breach', *The Guardian*, 17 March. https://www.theguardian.com/news/2018/mar/17/cambridge-analytica-facebook-influence-us-election (Accessed 7 April 2022).

Carr, N. (2016) *The Glass Cage: Who Needs Humans Anyway*. London: Vintage.

Center for the Governance of Change (2021) *European Tech Insights Report 2021*. https://www.ie.edu/cgc/research/european-tech-insights/ (Accessed 7 April 2022).

Chapman, S. (2021) 'Why the Bitcoin crash won't halt the growth in crypto assets', *New Statesman*, 7 July. https://www.newstatesman.com/politics/economy/2021/07/why-bitcoin-crash-won-t-halt-growth-crypto-assets (Accessed 7 April 2022).

Cheney-Lippold, J. (2011) 'A new algorithmic identity: Soft biopolitics and the modulation of control', *Theory, Culture & Society* 28: 164–81.

Cheney-Lippold, J. (2017) *We Are Data: Algorithms and the Making of Our Digital Selves*. New York: New York University Press.

Chomsky, N. (1971) 'The case against B.F. Skinner', *The New York Review of Books*, 30 December. https://www.ehu.eus/HEB/KEPA/The%20Case%20Against%20B.F.%20Skinner.pdf (Accessed 7 April 2022).

CNBC (2021) 'More than half of Europeans want to replace lawmakers with AI, study says', 27 May. https://www.cnbc.com/2021/05/27/europeans-want-to-replace-lawmakers-with-ai.html (Accessed 29 March 2022).

Cohen, D. (2014) *Homo Economicus: The Lost Profit of Modern Times*. Cambridge: Polity Press.

Colt, S. (2014) 'Tim Cook gave his most in-depth interview to date — Here's what he said', *Business Insider Australia*, 20 September. https://www.businessinsider.com/tim-cook-full-interview-with-charlie-rose-with-transcript-2014-9?r=US&IR=T (Accessed 7 April 2022).

Cooper, R. (2020) 'Pastoral power and algorithmic governmentality', *Theory, Culture and Society* 37: 29–52.

Crary, J. (2014) *24/7: Late Capitalism and the End of Sleep*. London: Verso.

Crawford, K., Lingel, J. and Karppi, T. (2015) 'Our metrics, our-selves: A hundred years of self-tracking from the weight scale to the wrist wearable device', *European Journal of Cultural Studies* 18: 479–96.

Curtis, A. (2011) *All Watched Over By Machines of Loving Grace* (BBC Film). https://www.filmsforaction.org/watch/bbc-all-watched-over-by-machines-of-loving-grace/ (Accessed 17 March 2022).

Darling, K. (2021) *The New Breed: How to Think About Robots*. Dublin: Allen Lane.

Davidson, J. (2016) 'Plenary address – A year of living "dangerously": Reflections on risk, trust, trauma and change', *Emotion, Space and Society* 18: 28–34.

de Dijn, A. (2020) *Freedom – An Unruly History*. Cambridge, MA: Harvard University Press.

Deleuze, G. (1990) *Postscript on the Societies of Control*. Anarchist Library.

Deloitte (2015) *Made to Order: The Rise of Mass Personalization*. https://www2.deloitte.com/ch/en/pages/consumer-business/articles/made-to-order-the-rise-of-mass-personalisation.html (Accessed 8 April 2022).

Dickson, E.J. (2019) 'Can Alexa and Facebook predict the end of your relationship?', *Vox*, January. https://www.vox.com/the-goods/2019/1/2/18159111/amazon-facebook-big-data-breakup-prediction (Accessed 2 March 2022).

Dow Schüll, N. (2016) 'Data for life: Wearable technology and the design of self-care', *BioSocieties*: 1–17. https://arts.mit.edu/wp-content/uploads/2016/11/Data-for-Life-Wearable-Technology-and-the-Design-of-Self-Care.pdf

Dunbar, R.I.M. (1992) 'Neocortex size as a constraint on group size in primates', *Journal of Human Evolution* 22: 469–93.

Eagle, N. and Pentland, A. (2006) 'Reality mining: Sensing complex social systems', *Personal and Ubiquitous Computing* 10: 255–68.

Edelman. (2021) *Edelman Trust Barometer 2021*. Edelman. https://www.edelman.com/trust/2022-trust-barometer (Accessed 25 August 2022).

Evans, D. (2011) *The Internet of Things – How the Next Evolution of Internet is Changing Everything*, Cisco ISBG, April. https://www.cisco.com/c/dam/en_us/about/ac79/docs/innov/IoT_IBSG_0411FINAL.pdf (Accessed 3 March 2022).

Fogg, B.J. (1998) 'Persuasive computers: Perspectives and research directions', *Conference on Human Factors in Computing Systems – Proceedings* (April): 225–32.

Foucault, M. (2008) *Security, Territory, Population – Lectures at the College de France, 1977–1978*. London: Palgrave MacMillan.

Fuchs, P., Dannenberg, D. and Wiedemann, C. (2021) 'Big Tech and labour resistance at Amazon', *Science as Culture* 31: 29–43.

Furedi, F. (2011) *On Tolerance: A Defence of Moral Independence*. London: Bloomsbury Continuum.

Gabrys, J. (2014) 'Programming environments: Environmentality and citizen sensing in the smart city', *Environment and Planning D: Society and Space* 32: 30–48.

Galloway, S. (2017) *The Four: The Hidden DNA of Amazon, Apple, Facebook and Google*. London: Transworld Publishers.

Galloway, S. (2021) *Post Corona: From Crisis to Opportunity*. London: Bantam Press.

Gladden, M.E. (2019) 'Who will be the members of society 5.0? Towards an anthropology of technologically posthumanized future societies', *Social Sciences* 8(5): 148.

Global Market Insights (2021) *Biometrics Market Size By Technology (Face, Fingerprint, Palmprint, Iris, Voice, Signature), By End-Use (BFSI, Government & Defense, Aerospace, Automotive, Consumer Electronics, Healthcare, Retail & E-Commerce), COVID-19 Impact Analysis, Regional Outlook.* https://www.gminsights.com/industry-analysis/biometrics-market (Accessed 18 April 2022).

Graeber, D. and Wengrow, D. (2021) *The Dawn of Everything: A New History of Humanity.* Dublin: Allen Lane.

Grand View Research (2021) *Wearable Technology Market Size, Share & Trends Analysis Report By Product (Wrist-Wear, Eye-Wear & Head-Wear, Foot-Wear, Neck-Wear, Body-Wear), By Application, By Region, And Segment Forecasts, 2021–2028.* https://www.grandviewresearch.com/industry-analysis/wearable-technology-market (Accessed 8 April 2022).

Grimmelmann, J. (2015) 'The virtues of moderation', *Yale Journal of Law and Technology* 17: 42–109.

Hao, K. (2021) 'How Facebook got addicted to spreading misinformation', *MIT Technological Review*, 11 March. https://www.technologyreview.com/2021/03/11/1020600/facebook-responsible-ai-misinformation/ (Accessed 8 April 2022).

Harari, Y.N. (2014) *Sapiens: A Brief History of Humankind.* London: Vintage Books.

Haraway, D. (1987) 'A manifesto for cyborgs: Science, technology, and socialist feminism in the 1960s', *Australian Feminist Studies* 4(Autumn): 1–42.

Haraway, D. (2004) *The Haraway Reader.* New York: Routledge.

Harvey, D. (1989) *The Urban Experience.* Baltimore: Johns Hopkins University Press.

Heath, T. (2016) 'This employee ID badge monitors and listens to you at work – except the bathroom', *Washington Post*, 7 September. https://www.washingtonpost.com/news/business/wp/2016/09/07/this-employee-badge-knows-not-only-where-you-are-but-whether-you-are-talking-to-your-co-workers/ (Accessed 12 August 2022).

Hern, A. (2019) 'Apple contractors "regularly hear confidential details" on Siri recordings', *The Guardian*, 26 July. https://www.theguardian.com/technology/2019/jul/26/apple-contractors-regularly-hear-confidential-details-on-siri-recordings (Accessed 8 April 2022).

Hill, K. (2019a) 'Life without the tech giants', *Gizmodo*, 22 January. https://gizmodo.com/life-without-the-tech-giants-1830258056 (Accessed 8 April 2022).

Hill, K. (2019b) 'I tried to block Amazon from my life. It was impossible', *Gizmodo*, 22 January. https://gizmodo.com/i-tried-to-block-amazon-from-my-life-it-was-impossible-1830565336 (Accessed 8 April 2022).

Hill, K. (2019c) 'I cut Google out of my life. It screwed-up everything', *Gizmodo*, 29 January. https://gizmodo.com/i-cut-google-out-of-my-life-it-screwed-up-everything-1830565500 (Accessed 8 April 2022).

Hill, K. (2019d) 'I cut Microsoft out of my life – or so I thought', 31 January. https://gizmodo.com/i-cut-microsoft-out-of-my-life-or-so-i-thought-1830863898 (Accessed 8 April 2022).

Hill, K (2019e) 'I cut Apple out of my life. It was devastating', *Gizmodo*, 2 February. https://gizmodo.com/i-cut-apple-out-of-my-life-it-was-devastating-1831063868 (Accessed 8 April 2022).

Hill, K. (2019f) 'I cut the big five tech giants from my life. It was hell', *Gizmodo*, 2 February. https://gizmodo.com/i-cut-the-big-five-tech-giants-from-my-life-it-was-hel-1831304194 (Accessed 8 April 2022).

Hill, K. (2019g) 'I cut Facebook out of my life. Surprisingly I missed it', *Gizmodo*, 24 January. https://gizmodo.com/i-cut-facebook-out-of-my-life-surprisingly-i-missed-i-1830565456 (Accessed 8 April 2022).

Hilts, A., Parsons, C. and Knocke, J. (2016) *Every Step you Fake: A Comparative Analysis of Fitness Tracker Privacy and Security – Open Effect Report*. https://openeffect.ca/reports/Every_Step_You_Fake.pdf (Accessed 7 April 2022).

House of Lords (2011) *Behaviour Change*. London: The Stationery Office.

Housing and Development Board (2014) *Smart HDB Homes of the Future*, 11 September. https://www.smartnation.gov.sg/media-hub/press-releases/smart-hdb-homes-of-the-future (Accessed 8 April 2022).

Howes, S. and Jones, K.M. (2019) 'COMPUTER SAYS "NO!" Stage one: Information provision'. https://cpag.org.uk/policy-and-campaigns/report/computer-says-no-stage-one-information-provision (Accessed 25 August 2022).

iSelect (n.d.) *5G vs. 4G*. https://www.iselect.com.au/internet/5g-australia/5g-vs-4g/ (Accessed 18 April 2022).

Isin, E. (2004) 'The neurotic citizen', *Citizenship Studies* 8: 217–35.

Isin, E. and Ruppert, E. (2020) 'The birth of sensory power: How a pandemic made it visible?', *Big Data and Society* (July–December): 1–15.

James, H. (1884) 'The art of fiction', *London Magazine*, 9 September. https://public.wsu.edu/~campbelld/amlit/artfiction.html (Accessed 18 April 2022).

Jeffries, A. (2011) 'What is digital veganism? Cody Brown explains his catchphrase', *Observer*, 28 June. https://observer.com/2011/06/what-is-digital-veganism-cody-brown-explains-his-catchphrase/ (Accessed 18 April 2022).

Jones, R. and Whitehead, M. (2018) '"Politics done like science": Critical perspectives on psychological governance and the experimental state', *Environment and Planning D: Society and Space* 36(2): 313–30.

Jones, R., Pykett, J. and Whitehead, M. (2011) 'Governing temptation: Changing behaviour in an age of libertarian paternalism', *Progress in Human Geography* 35: 483–501.

Jones, R., Pykett, J. and Whitehead, M. (2013) *Changing Behaviours: On the Rise of the Psychological State*. Cheltenham, UK, and Northampton, MA, USA: Edward Elgar Publishing.

Kahneman, D. (2012) *Thinking, Fast and Slow*. London: Penguin.

Kahneman, D. and Klein, G. (2009) 'Conditions for intuitive expertise: A failure to disagree', *American Psychologist* 64: 515–26.

Keen, A. (2015) *The Internet Is Not the Answer*. London: Atlantic Books.

Kegan, R. (1982) *The Evolving Self: Problem And Process In Human Development*. Cambridge, MA: Harvard University Press.

Kennedy, H., Poell, T. and van Dijck, J. (2015) 'Data and agency', *Big Data and Society* 2: 1–7.

Kitchin, R. (2015) 'Making sense of smart cities: Addressing present shortcomings', *Cambridge Journal of Regions, Economy and Society* 8(1): 131–6.

Kitchin, R. and Fraser, A. (2020) *Slow Computing: Why We Need Balanced Digital Lives*. Bristol: Bristol University Press.

Knox, J., Williamson, B. and Bayne, S. (2020) 'Machine behaviourism: Future visions of "learnification" and "datafication" across humans and digital technologies', *Learning, Media and Technology* 45: 31–45.

Kurtz, C.F. and Snowden, D.J. (2003) 'The new dynamics of strategy: Sense-making in a complex and complicated world', *IEEE Engineering Management Review* 31: 462–483.

Lago, C. (2021) 'Is it too soon to automate policy: The use of AI in government must be fair, transparent and accountable', *The New Statesman*, 17 September. https://www

.newstatesman.com/spotlight/2021/09/is-it-too-soon-to-automate-policy (Accessed 12 August 2022).

Lanier, J. (2018) *Ten Arguments for Deleting Your Social Media Accounts Right Now*. London: The Bodley Head.

Lanzing, M. (2019) '"Strongly recommended" revisiting decisional privacy to judge hypernudging in self-tracking technologies', *Philosophy and Technology* 32: 549–68.

Law Commission (2022) *Automated Vehicles: Summary of Joint Report*. London: Law Commission.

Legget, W. (2014) 'The politics of behaviour change: Nudge, neoliberalism and the state', *Policy and Politics* 42: 3–19.

Leslie, D. (2019) 'Understanding artificial intelligence ethics and safety: A guide for the responsible design and implementation of AI systems in the public sector', *The Alan Turing Institute*. https://www.turing.ac.uk/sites/default/files/2019-06/understanding _artificial_intelligence_ethics_and_safety.pdf (Accessed 19 April 2022).

Lovelock, J. (2019) *The Novocene: The Coming Age of Hyperintelligence*. London: Penguin.

Low, D. (2021) 'Singapore is the world's smartest city for the third year: IDM Smart City Index', *The Straits Times*, 2 November. https://www.straitstimes.com/tech/tech -news/singapore-is-worlds-smartest-city-for-the-third-year-imd-smart-city-index (Accessed 19 April 2022).

Lupton, D. (2016) *The Quantified Self: A Sociology of Self-Tracking*. Cambridge: Polity Press.

Lyons, E.J., Lewis, Z.H., Mayrsohn, B.G. and Rowland, J. (2014) 'Behavior change techniques implemented in electronic lifestyle activity monitors: A systematic content analysis', *Journal of Medical Internet Research* 16(8): e192.

Maçães, B. (2021) 'The spirit of the age: Why the tech billionaires want to leave human-ity behind Jeff Bezos and his contemporaries', *The New Statesman*, 29 November. https://www.newstatesman.com/science-tech/2021/09/the-spirit-of-the-age-why-the -tech-billionaires-want-to-leave-humanity-behind (Accessed 12 August 2022).

Matisse, N. (2015) 'Robot depending on kindness of strangers meets its demise in Philadelphia', *ars Techica*. https://arstechnica.com/gaming/2015/08/robot -depending-on-kindness-of-strangers-meets-its-demise-in-philadelphia/ (Accessed 19 April 2022).

Mattern, S. (2013) 'Methodolatry and art of measure: The new wave of urban data science', *Places Journal*, November. https://placesjournal.org/article/methodolatry -and-the-art-of-measure/?cn-reloaded=1 (Accessed 19 April 2022).

Mattern, S. (2016) 'Instrumental city: The view from Hudson Yards, circa 2019', *Places Journal*, April. https://placesjournal.org/article/instrumental-city-new-york -hudson-yards/?cn-reloaded=1 (Accessed 19 April 2022).

McDaniel, B.T. and Coyne, S.M. (2016) '"Technoference": The interference of tech-nology in couple relationships and implications for women's personal and relational well-being', *Psychology of Popular Media Culture* 5: 85–98.

McDonald, A.M. and Cranor, L.F. (2008) 'The cost of reading privacy policies', *A Journal of Law and Policy for the Information Society* 4. https://lorrie.cranor.org/ pubs/readingPolicyCost-authorDraft.pdf (Accessed 19 April 2022).

McNamee, R. (2020) *Zucked: Waking Up to the Facebook Catastrophe*. London: Harper Collins.

Monahan, T., Phillips, D.J. and Wood, D.M. (2010) 'Surveillance and empowerment', *Surveillance and Society* 8: 106–12.

Munger, K. (2021) 'Hello goodbye: The more conversations machines produce for us, the harder it will become to say something non algorithmic', *Real Life*, 12 April. https://reallifemag.com/hello-goodbye/ (Accessed 19 April 2022).

Musk, E. (2017) *World Government Summit 2017: A Conversation with Elon Musk*. https://www.youtube.com/watch?v=Xa8m3SATR1s (Accessed 19 April 2022).

Neate, R. (2018) 'Over $119bn wiped off Facebook's market cap after growth shock', *The Guardian*, 26 July. https://www.theguardian.com/technology/2018/jul/26/facebook-market-cap-falls-109bn-dollars-after-growth-shock (Accessed 12 August 2022).

O'Connor, S. (2021) 'Why I was wrong to be optimistic about robots: Humans are being crunched into a robot system working at a robot pace', *Financial Times*, 8 February. https://www.ft.com/content/087fce16-3924-4348-8390-235b435c53b2 (Accessed 12 August 2022).

O'Grady, F. (2021) 'The trade union's role in automation: AI has tremendous potential in the workplace – but we need to make sure we put people first', *The New Statesman*, 17 September. https://www.newstatesman.com/spotlight/2021/09/automation-and-the-trade-unions (Accessed 12 August 2022).

Ofcom (2019) *Adults' Media Lives 2019*. London: OFCOM.

Office for National Statistics (ONS) (2016) 'How is the welfare budget spent?', 16 March. https://www.ons.gov.uk/economy/governmentpublicsectorandtaxes/publicsectorfinance/articles/howisthewelfarebudgetspent/2016-03-16 (Accessed 19 April 2022).

Ohanian, A. (2013) *Without Their Permission: How the 21st Century Will be Made Not Managed*. New York: Grand Central Publishing.

Olguín, D.O., Waber, B.N., Kim, T., Mohan, A., Ara, K. and Pentland, A. (2009) 'Sensible organizations: Technology and methodology for automatically measuring organizational behavior', *IEEE Transactions on Systems, Man, and Cybernetics, Part B: Cybernetics* 39: 43–55.

Ollila, E. (2018) 'Will smart badges help employees strategize?', *Spark*. https://www.adp.com/spark/articles/2018/04/will-smart-badges-help-employers-strategize.aspx (Accessed 20 October 2021).

Osnos, E. (2018) 'Can Mark Zuckerberg fix Facebook before it breaks democracy?', *The New Yorker*, 10 September. https://www.newyorker.com/magazine/2018/09/17/can-mark-zuckerberg-fix-facebook-before-it-breaks-democracy (Accessed 19 April 2022).

Pasquale, F. (2015). *The Black Box Society*. Cambridge, MA: Harvard University Press.

Pasquale, F. and Bracha, O. (2015) 'Federal Search Commission? Access, fairness and accountability in the law of search', *Cornell Law Review* 93: 1149–91.

Pasquale, J. (2020) *New Laws of Robotics: Redefining Human Expertise in the Age of AI*. Cambridge, MA: Belknap Press.

Pentland, A. (2014) *Social Physics: How Good Ideas Spread – The Lessons from a New Science*. London: Penguin Books.

Perrin, A. (2018) 'Americans are changing their relationship with Facebook', Pew Research Center, 5 September. https://www.pewresearch.org/fact-tank/2018/09/05/americans-are-changing-their-relationship-with-facebook/ (Accessed 19 April 2022).

Phillips, A. (2012) *Missing Out: In Praise of the Unlived Life*. London: Penguin Books

Pilkington, E. (2019) 'Digital dystopia: How algorithms punish the poor', *The Guardian*, 14 October. https://www.theguardian.com/technology/2019/oct/14/automating-poverty-algorithms-punish-poor (Accessed 12 August 2022).

Pound, N. (2020) *Challenging Amazon: What can we do about Amazon's Treatment of its Workers*. London: Trades Union Congress.

Rashid, I. and Kenner, S. (2019) *Offline: Free Your Mind from Smartphone and Social Media Stress*. Chichester: John Wiley & Sons.

Risdon, C. (2017) 'Scaling nudges with machine learning', *Behavioral Scientist*, 25 October. http://www.behavioralscientist.org/scaling-nudges-machine-learning/ (Accessed 19 April 2022).

Rose, N. (1998) *Inventing Ourselves: Psychology, Power, and Personhood*. Cambridge: Cambridge University Press.

Schneider, N. (2015) 'The joy of slow computing', *The New Republic*, 20 May. https://newrepublic.com/article/121832/pleasure-do-it-yourself-slow-computing (Accessed 19 April 2022).

Scott, J. (1999) *Seeing Like a State: How Certain Schemes to Improve the Human Condition Have Failed*. New Haven, CT: Yale University Press.

Skinner, B.F. (1972) *Beyond Freedom and Dignity*. London: Penguin.

Smart Nation and Digital Government Office (2018) *Smart Nation: The Way Forward*. https://www.smartnation.gov.sg/files/publications/smart-nation-strategy-nov2018.pdf (Accessed 19 April 2022).

Smart Nation Singapore (2022) *Elderly Monitoring System*. https://www.smartnation.gov.sg/initiatives/urban-living/ems (Accessed 19 April 2022).

Smith, D.W. (2018) 'Phenomenology', in *The Stanford Encyclopedia of Philosophy*, Zalta, E.N. (ed.). https://plato.stanford.edu/archives/sum2018/entries/phenomenology/ (Accessed 19 April 2022).

Smith, K. (2019) '53 incredible Facebook statistics and facts', *Brandwatch*. https://www.brandwatch.com/blog/facebook-statistics/ (Accessed 19 April 2022).

Snowden, D. (2002) 'Complex acts of knowing: Paradox and descriptive self-awareness', *Journal of Knowledge Management* 6(2): 100–11.

Stamos, A. (2019) 'The platform challenge: Balancing safety, privacy and freedom', DataEdge Berkley School of Information. https://www.youtube.com/watch?v=ATmQj787Jcc (from 10.21 minutes) (Accessed 19 April 2022).

Strategyr (2022) *Smart Cities World Market Report*. https://www.strategyr.com/market-report-smart-cities-forecasts-global-industry-analysts-inc.asp (Accessed 19 April 2022).

Strengers, Y. and Kennedy, J. (2020) *The Smart Wife: Why Siri, Alexa and Other Smart Home Devices Need a Feminist Reboot*. Cambridge, MA: MIT Press.

Sunstein, C. (2016) *The Ethics of Influence: Government in the Age of Behavioural Science*. Cambridge: Cambridge University Press.

Sunstein, C. (2019) *On Freedom*. Princeton, NJ: Princeton University Press.

Taylor, A. (2018) 'The automation charade', *Logic*, August. https://logicmag.io/failure/the-automation-charade/ (Accessed 19 April 2022).

Thaler, R. (2015) *Misbehaving: The Making of Behavioural Economics*. London: Penguin.

Thaler, R. and Sunstein, C. (2009) *Nudge: Improving Decisions About Wealth, Health and Happiness*. London: Penguin.

Uber (n.d.) *How Surge Pricing Works*. https://www.uber.com/gb/en/drive/driver-app/how-surge-works/ (Accessed 19 April 2022).

UK Government (2016) *Government Transformation Strategy*. London: Cabinet Office.

United Nations Human Rights Council (2019) *General Assembly, Report of the Special Rapporteur on Extreme Poverty and Human Rights*. New York: United Nations.

van Dijck, J. (2014) 'Datafication, dataism and dataveillance: Big data between scientific paradigm and ideology', *Surveillance and Society* 12: 197–208.

Vanolo, A. (2014) 'Smartmentality: The smart city as disciplinary strategy', *Urban Studies* 51: 883–98.

Wallace-Stephen, F. and Morgante, E. (2020) *Who Is at Risk? Work and Automation, in the Time of Covid-19*. London: RSA.

Warrell, H. (2019) 'Home Office under fire for using secretive visa algorithm', *Financial Times*, 9 June. https://www.ft.com/content/0206dd56-87b0-11e9-a028 -86cea8523dc2 (Accessed 12 August 2022).

Waterson, J. (2020) 'Boris Johnson urges top UK tech firms to join coronavirus fight', *The Guardian*, 13 March. https://www.theguardian.com/business/2020/mar/ 13/johnson-urges-top-uk-tech-firms-to-join-coronavirus-fight (Accessed 19 April 2022).

Weick, K., Sutcliffe, K.M. and Obstfeld, D. (2005) 'Organising and the process of sensemaking', *Organization Science* 16: 409–21.

Weinmann, M., Schneider, C. and vom Brocke, J. (2016) 'Digital nudging', *Business and Information Systems Engineering* 58: 433–6.

Whitehead, A.N. (1911) *An Introduction to Mathematics*. Ann Arbor: University of Michigan Library.

Whitehead, M. (2020) 'Why people leave Facebook – and what it tells us about the future of social media', *The Conversation*. https://theconversation.com/amp/ why-people-leave-facebook-and-what-it-tells-us-about-the-future-of-social-media -128952 (Accessed 19 April 2022).

Whitehead, M., Jones, R., Lilley, R., Pykett, J. and Howell, R. (2017) *Neuroliberalism: Behavioural Government in the Twenty-First Century*. Abingdon: Routledge.

Whitehead, M., Jones, R. and Pykett, J. (2011) 'Governing irrationality, or a more than rational government? Reflections on the rescientisation of decision making in British public policy', *Environment and Planning A* 43: 2819–37.

Wired Staff (2000) 'Fiddling with human behavior', *Wired*. https://www.wired.com/ 2000/03/fiddling-with-human-behavior/ (Accessed 12 September 2022).

Wired Staff (2009) 'Know thyself: Tracking every facet of life, from sleep to mood to pain, 24/7/365', *Wired*. https://www.wired.com/2009/06/lbnp-knowthyself/ (Accessed 19 April 2022).

Wood, D.M. and Wright, S. (2015) 'Before and after Snowden', *Surveillance and Society* 13: 132–8.

World Economic Forum (2020a) *The Future of Jobs Report 2020*, October. https:// www.weforum.org/reports/the-future-of-jobs-report-2020/ (Accessed 12 August 2022).

World Economic Forum (2020b) 'The tiny nation is showing the world how technology can make a huge difference to the elderly', 17 January. https://www.weforum.org/ agenda/2020/01/ageing-demographic-elderly-technology-singapore/ (Accessed 19 April 2022).

Yeo, S.J.I. (2022) 'Smart urban living in Singapore? Thinking through everyday geographies', *Urban Geography*. https://doi.org/10.1080/02723638.2021.2016258.

Yeung, K. (2016) '"Hypernudge": Big Data as a mode of regulation by design', *Information, Communication and Society* 20: 118–36.

Završnik, A. (2021) 'Algorithmic justice: Algorithms and big data in criminal justice settings', *European Journal of Criminology* 18: 623–42.

Zuboff, S. (2019) *The Age of Surveillance Capitalism: The Fight for a Human Future at the New Frontier of Power*. London: Profile Books.

Index

5G technology 146
1818 *Frankenstein* (Shelley) 9
1972 *Beyond Freedom and Dignity*
    (Skinner) 94
2012 Government Digital Strategy 159
2014 Smart Nation programme 167

acceleration 182
Age of Enlightenment 151
Agricultural Revolution 1
Akerloff, G. 77, 78
algorithmic government 152
Alston, P. 160
Amazon 176
Amazon fulfilment centre 133, 138
Amazon's fulfilment centres 136, 146
American Psychologist 82
Amoore, L. 30, 44, 45, 91, 152, 164
Anderson, W. C. 15, 16, 17
Animal Spirits: How Human Psychology
    Drives the Economy, and
    Why it Matters for Global
    Capitalism 77
An Introduction to Mathematics 3
Anthropocene 13
anthropological monster 77
anticipatory ethics 35
Apple 178
Apple Watch 122
applied utopianism 8
artificial intelligence (AI) 19
augmentation 129
Automated Vehicles Act 202
automation 18
auto-technological ethnography 100

Bailey, P. 194
Ball, J. 27
Bank of America 127
Bank of England 130, 131
Barbrook, R. 12
Barlow, J.P. 11, 12

Bastani, A. 1, 11
Bedtime Consistency Score 117
Beer, D. 27, 29, 30, 31, 182
Behavioral Scientist (2017) 87
behaviour 75
Behaviour Design Lab 97
behavioural actuation 76, 94, 95, 122
behavioural conditioning 94, 96
behavioural economics 78, 80
behavioural economists 78, 79, 82
behavioural prediction 45
behavioural science 87
behavioural techniques 106
behaviouralist ethos 110
behaviouralists 75
behaviours 75
Berreby, D. 22, 23, 33, 34, 35
Big-Tech 128
    benefits of the Covid-19 pandemic 148
biometrics technologies 101
Blade Runner (Scott) 9
bold social experiments 1
bounded rationality 78
Bourdieu, P. 77

Californian Ideology 12
Cambridge Analytica scandal 192, 193
Cameron, A. 8, 12
Cameron, J. 9
Carr, N. 20, 23, 24
CEO 142
Challenging Amazon 134
Chapman, S. 98
Cheney-Lippold, J. 22, 43, 44, 60, 73,
    74, 99, 111, 114
Chief Knowledge Officer 140
Child Poverty Action Group 164, 165
choice architectures 80
civilization 3
Cloud Ethics (Amoore) 44
cognitive bias 78
cognitive heuristics 79

Computer says 'No!' 164
Conner, S. O. 24, 25
Conservative Government 160
Cook, T. 107
Cooper, R. 152, 153
Corporate Social Responsibility 140
counterfeiting of humanity 33
Covid-19 pandemic 128, 148, 155, 165
Coyne, S. M. 174
Crannor, L. F. 92
Crary, J. 113, 118
Crawford, K. 102, 105, 110, 112, 124
Credit Crunch 78
creepy 61
cultivation 5
Cyberdyne Systems 8
cyborgs 25, 41, 42

dark factories 130
Darling, K. 130, 137, 138
data bias 30
data boosterism 29
data gaze 29, 30, 31
data-informed knowledge 27, 28, 29, 31
data rationality 31
data science 87
data self 41, 43, 73
data selves 99
data sovereignty 183, 184
dataveillance 38, 136
Davidson, J. 99, 110, 111, 112, 116, 124, 125
Declaration of the Independence of Cyber Space 11
Defoe, D. 77
de-Googling 177
dehumanised intensification 25
Deleuze, G. 154, 155
Deloitte 46
democratic membership organization 181
dialectical learning processes 3
Dick, P.K. 9
digital behaviours 76, 98
digital communication 22
digital data surveillance 147
digital data transfer 145
digital disconnection 173
digital Dunkirk 148
digital realm 12

digital self-tracking 109
digital Taylorism 134
digital technology 1
digital veganism 173, 180
digital workplace monitoring 129
disruption
    period of 1
Distance Assistance technology 135
Do Androids Dream of Electric Sheep? 9
dumbing down 173, 201
Dunbar, R. 196

economic human 76
economies of action 93
economists 186
Edelman 142
Emotional Contagion Trial 90
Every Step you Fake 107
extension 47
extraction 182

Facebook 11, 46, 59, 67, 128, 142, 177
    digital ecosystem 200
    user data 192
Facebook deleters 173, 174, 177, 185, 188
fauxtomation 22, 23
First Disruption 5
flow room 134
Fogg, B.J. 97
Foucauldian approaches 39
Foucauldian theories of 152
Foucault, M. 79, 150, 151, 152, 153, 154, 155, 159
Fraser, A. 182, 183, 184
freedom
    choice 86
    negative 85
    positive 85
Fuchs, P. 138
Fully Automated Luxury Communism (Bastani) 1

Gillmor, D.K. 179
Global Market Insights 101
Google 11, 46, 68
Government Transformation Strategy 159, 160
governmental intervention 157
governmentality 151
    algorithmic 152

Foucauldian theories of 152, 155
liberal forms of 151
model for 152
Great Recession 77, 78
Grimmelmann, J. 142
Gummer, B. 159

Harari, Y. N. 5, 6
Haraway, D. 36, 37, 106, 107
Harvey, D. 167
herd bias 79
herding 94
Hill, K. 175, 176, 177, 178, 179, 180,
    188, 191, 194, 200
HitchBOT 34, 35
homo economicus 76, 77, 92
human behaviour
  action 75
  activities 75
  homo economicus 76
  neoliberal model 78
  predictability 95
  prediction 95
human behaviour 75
human displacement 23
human mimicry 33
human–smart-tech interface
  processes affected vi, 19
  properties 19
humane automation 26
hypernudges 83, 84, 85, 86, 87, 88, 89,
    90, 91, 92, 93, 95, 98
  negative freedom 85
  positive freedom 85
  transparency in 87

iLoss 178
imaginative resource 107
Instagram 63
instrumentarian power 39
instrumentarianism 153
International Center for Media & the
    Public Agenda (ICMPA) 199
internet-enabled fabrics 100
Isin, E. 111, 155, 156

Japan 5.0 171
Johnson, B. 148

Kahneman, Daniel 78, 79, 82
Kelly, K. 104

Kenner, S. 174
Kitchin, R. 166, 182, 183, 184
Knox, J. 205
Kurtz, C. F. 50, 51

Lago 157
Lanier, J. 174, 175
Law Commission 202, 203
  principle 203
laws of robotics 8
Learning and intelligence 206
libertarian paternalism 86
Lovelock, J. 13, 14
Lupton, D. 104, 106, 107, 108, 109, 110,
    112, 124
luxury trap 6, 7
Lyons, E.J. 106

machinic communication 20
Mackay, R. S. 108
Maddern, S. 172
Manifesto for Cyborgs 36
Manifesto for Cyborgs (Haraway) 37
McDaniel, B.T. 174
McDonald, A.M. 92
McDonald's 144, 145
McLuhan, M. 12
mentality of government 151
micronudges 83
Microsoft 178
*Missing Out: In Praise of the Unlived
    Life* 207
Munger, K. 20, 21, 22
Musk, E. 25

National Computerization 167
neoliberalism 12, 77
neoliberals 78
Neolithic period 1, 6
Neolithic Revolution 1, 5
neurotic citizens 111
notice and consent system(s) 91, 92
Novocene 13
*Nudge: Improving Decisions
    About Health, Wealth and
    Happiness* 79
nudges 80

O'Connor, S.O. 135
Office for National Statistics 158

partial reinforcement 74
Pasquale, J. 25, 32, 33, 34, 35, 138, 139
Pentland, A. 135
personalization 46
'Persuasive Computers: Perspectives and
    Research Directions' (Fogg) 97
pessimistic theories 2
Petabyte Age 15, 16
Pew Research Centre's American Trends
    Panel 193
Phillips, A. 207
Pilkington, E. 159
Plato 150
Postscript on Societies of Control
    (Deleuze) 154
prediction 43
  digital 57
  potential 65
  search 62
prediction imperative 46, 73, 76
prediction paradox 73, 74
prediction products 46, 73
predictive paradox 64, 72, 205
predictive products 74
professionalism 138

Quantified-Self Movement 103, 104,
    105, 112

radical technologies 102
Rand, A. 11
Rashid, I. 174
rationality of speed 182
reasonable discernment 87
relationship outcome prediction
    technologies 3
Risdon, C. 87, 88, 89
robot colleagues 25
robots 19, 32
Rousseau 150
Ruppert, E. 155, 156

Schneider, N. 181, 182, 184
Scott, A. 151
Scott, R. 9
self-awareness 3
self-monitoring 109, 125
self-quantification 114, 125, 206
Self-Quantification Movement 120, 121
SenseMaker
  approach 49, 50

method 49
narrative 60
narratives 61, 67
survey 53, 56, 62
surveys 51
SenseMaker survey 185, 189
sensory power 155
Shelley, M. 9, 10
Shiller, R. 77, 78
Simon, H. 78
Singapore's Elderly Monitoring
    System 170
Skinner, B.F. 94, 95, 96
SLAM (Scan, Label, Apply,
    Manifest) 134
Slow Computing 173, 180, 181, 182,
    183, 184, 199, 200, 201, 204
smart body 99
  origins of 108
smart-body technology 206
smart cities 165
smart city
  agenda 166
smart governmentality 149, 151, 152,
    153, 154, 156, 158, 160, 163, 170
Smart Home initiative 169
Smart Neighbourhood programme 169,
    170
smart pedagogy 205
smart-tech blocks 176
smart-tech firms 141
smart technologies 2
  capabilities of 2
  human interface 2
  impact on 4
  reassured by 64
  revolution 11
smart technology 19, 103
  advantage of 18
  and corporation 127
  anthropomorphising 23
  augmentation 129
  monopolies 180
  nudges 82
  optimal performance of 158
  prediction 47
  predictive aspects of 43
smart-tech predictions
  negative implications of 71
smart urbanism 167

smart warehouses 24
smart workplaces 128, 130, 131, 135, 136, 137
smartmentality 171
Smartwatch 112
Snowden, D. 49, 50, 51
Snowden, E. 181
social communication 20, 21
social comparison 198
social disconnection
    problems of 200
    processes of 200
social influence 79
social media platform 186, 198
social media platforms 65, 71, 185, 188
    benefits of 186
    facilitates 187
    impacts of 186
    predictive power of 69, 70
social media predictions 72, 73
Society of Control 154, 155
socio-digital re-orientation 182
socio-economic problems 80
socio-technological phenomenology 40, 41, 42
Stamos, A. 127, 128
Stanford Persuasive Technology Lab 97
status quo bias 79
Sunstein, C. 78
surveillance 136
surveillance capitalism 24, 38, 39, 41, 70, 73, 97
Surveillance Capitalism 121
surveillance capitalists 96

talk-activated systems 68
Taylor, A. 22, 23
tech-free sensing self 99
technoference 174
technological integration 130
techno-phenomenological approach 50
Ten Arguments for Deleting Your Social Media Accounts Right Now 174
Terminator (Cameron) 9
Thaler, R. 78
The Data Gaze: Capitalism Power and Perception 29
'The End of Theory: The Data Deluge Makes the Scientific Method Obsolete' (Anderson) 15

The Fountainhead and *Atlas Shrugged* (Rand) 11
The Matrix 10
The Minority Report (Dick) 9, 10
The New Laws of Robotics (Pasquale) 25
The Novocene: The Coming Age of Hyperintelligence 13
The *Observer* 192
Third Disruption 1, 4, 5, 6, 7, 10, 13
Thoreau 179
Trades Union Congress (TUC) 134
transition demands 202, 203
transition period 203
Trump, Donald 173
Tumblr 63
tuning 93
Turing, A. 44
Tversky, A. 78
Twitter 128
    algorithm 58

Uber 143, 144
UK's Digital Dunkirk 149
UK's Universal Credit System 161
UK's Universal Credit Systems 163
UN Human Rights Council (2019) 163
United Nations Humans Rights Council 162
Universal Credit System 162
Utilities Management Systems 170
utopianism 8, 11

van Dijck, J. 38
Vanolo, A. 171
van Veen, C. 160
Voter Megaphone project 83, 84

We Are Data (Cheney-Lippold) 180
wearable digital technology 109, 112
wearable technologies 105
WhatsApp 68
Whitehead, A.N. 3
Wired 104
Wittgenstein, L. 44
Wolf, G. 104
Word Cloud 56, 57
World Economic Forum 130
World Economic Forum's *Future of Job Report 2020* 137
World Government Summit 25

Yeo, S.J.I. 168, 169, 170, 171
Yeung, K. 83, 89, 90, 91, 92, 93
Yuhua Estate 168

zero-click ordering 62
zombie-debt 163

Zuboff, S. 8, 16, 17, 24, 27, 38, 39, 40,
    45, 46, 47, 61, 70, 73, 74, 76, 93,
    94, 95, 96, 97, 98, 100, 107, 108,
    109, 126, 153, 205
Zuckerberg, M. 173, 193

Printed and bound by CPI Group (UK) Ltd, Croydon, CR0 4YY

27/10/2024

14580411-0002